Nelson International Mathematics
Student Book 4

2nd edition

OXFORD UNIVERSITY PRESS

OXFORD
UNIVERSITY PRESS

Great Clarendon Street, Oxford, OX2 6DP, United Kingdom

Oxford University Press is a department of the University of Oxford.
It furthers the University's objective of excellence in research, scholarship,
and education by publishing worldwide. Oxford is a registered trade mark of
Oxford University Press in the UK and in certain other countries

Text © Cloud Publishing Services 2013
Original illustrations © Oxford University Press 2014

The moral rights of the authors have been asserted

First published by Nelson Thornes Ltd in 2013
This edition published by Oxford University Press in 2014

All rights reserved. No part of this publication may be reproduced,
stored in a retrieval system, or transmitted, in any form or by any
means, without the prior permission in writing of Oxford University
Press, or as expressly permitted by law, by licence or under terms
agreed with the appropriate reprographics rights organization.
Enquiries concerning reproduction outside the scope of the above
should be sent to the Rights Department, Oxford University Press, at
the address above.

You must not circulate this work in any other form and you must
impose this same condition on any acquirer

British Library Cataloguing in Publication Data
Data available

978-1-4085-1903-5

11

Printed and bound by CPI Group (UK) Ltd, Croydon, CR0 4YY

Acknowledgements

Cover artwork: Andy Peters
Illustrations: Alan Rogers, Pantek Media, Maidstone and OKS Prepress, India
Page make-up: Pantek Media, Maidstone

Photographs
iStockphoto: pp47, 48, 53, 115

Although we have made every effort to trace and contact all
copyright holders before publication this has not been possible in all
cases. If notified, the publisher will rectify any errors or omissions at
the earliest opportunity.

Links to third party websites are provided by Oxford in good faith
and for information only. Oxford disclaims any responsibility for
the materials contained in any third party website referenced in
this work.

Contents

Revising place value	5	
Place value to thousands	6	
Working with larger numbers	7	
Comparing and ordering numbers	8	
Compare and order using < and >	9	
More ordering and comparing numbers	10	
Rounding to the nearest 10	11	
Rounding to the nearest 100	12	
Revising 2D shapes	13	
Investigating polygons	14	
Pinboard investigations	15	
More about quadrilaterals	16	
Investigating rectangles	17	
Analogue time	18	
Time: a.m. and p.m.	19	
Digital time	20	
Make your own sliding digital clock	21	
Using a calendar	22	
Timetables	23	
Decimal notation	24	
Decimal place value	25	
Counting in tenths	26	
Decimals and fractions	27	
More decimals	28	
Comparing decimals	29	
Gymnastics scores	30	
Metric units of length	31	
Reading and writing lengths	32	
Litres and millilitres	33	
Mixing amounts	34	
Using liquids	35	

Which is best?	36	
Reading measuring scales	37	
Counting on and back	38	
Adding or subtracting groups of 10s, 100s and 1000s	39	
Making 100	40	
Pairs of numbers that make 1000	41	
Revising addition facts	42	
Adding numbers by making 10 or 20	43	
Adding multiples of 10	44	
Symmetry	45	
Symmetry in polygons	46	
Symmetry around us	47	
Organising data	48	
Frequency tables	49	
A typical day	50	
Who uses the shop?	51	
More bar charts	52	
Reading a pictogram	53	
More pictograms	54	
Drawing pictograms	55	
Revising repeated addition and subtraction	56	
Revising the 2×, 3×, 5× and 10× tables	57	
The 4× table	58	
The 6× and 9× tables	59	
How many ways?	60	
Multiplication grids	61	
Code breakers	62	
Division facts	63	
Revising 3D shapes	64	
Naming 3D shapes	65	

Contents

Solids and their nets	66
Is it a cuboid?	67
Is it a pyramid?	68
Negative numbers	69
Reading a thermometer	70
One day in winter	71
Fractions	72
Fractions of shapes	73
Fractions of a number	74
Comparing fractions	75
Equivalent fractions	76
Compare and order equivalent fractions	77
Fractions and decimals	78
Mixed numbers	79
Real-life problems	80
Position on a grid	81
Compass directions	82
Finding your way	83
Mental strategies for adding	84
Estimating	85
Counting on and back to subtract	86
More subtraction strategies	87
Working with bigger numbers	88
More adding and subtracting	89
Coded subtractions	90
Fruit and nut problems	91
Perimeter	92
Area	93
More area	94
Odd and even numbers	95
Zig-zag number track	96
Number patterning	97
Multiples	98
Angles	99
Compare and order angles	100
Revise multiplication facts	101
Multiplying tens	102
Multiply bigger numbers by 10	103
Multiply by 100	104
Doubling	105
Halving	106
Multiplying a two-digit number by a one-digit number	107
Multiplication problems	108
Venn diagrams	109
More Venn diagrams	110
Caroll diagrams	111
Sorting data into three groups	112
Sorting data into three groups *continued*	113
Using Carroll diagrams to sort data	114
A database: big cats	115
Using a database	116
More sorting	117
Dividing by sharing	118
Division by repeated subtraction	119
More dividing	120
Rounding answers after division	121
Dividing by 10 and 100	122
Divide or multiply?	123
Ratio and proportion	124
Ratio and proportion problems	125
Classroom proportions	126
Glossary	127

Revising place value

Look at this number: 625
We say six hundred and twenty five.
We can show this number on a place value table like this:

Hundreds	Tens	Units
6	2	5

We can partition the number and write it in expanded form like this:
$625 = (6 \times 100) + (2 \times 10) + (5 \times 1)$
$= 600 + 20 + 5$

1 Count and work out the number. Write each number using expanded form.

a How many bags of crisps altogether?

b How many mints altogether?

2 Write the numbers shown by these place value cards.

a 100, 100, 10, 1, 100, 10, 10, 1, 1, 100, 10, 10, 1

b 100, 10, 10, 1, 10, 10, 10, 10, 1, 1, 10, 10, 10, 1, 1

c 100, 1, 1, 1, 1, 100, 1, 1, 1

d 100, 100, 100, 1, 100, 100, 10, 10, 100, 100, 10

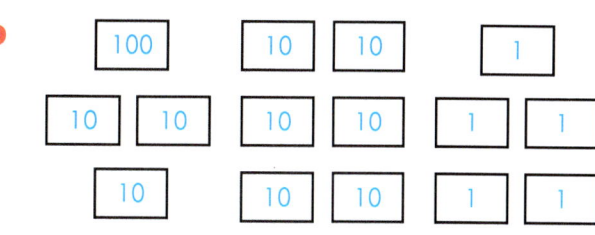

3 Say each number. Then write it in expanded form.

a 613   b 226   c 879   d 547
e 984   f 732   g 461   h 358

you can use Workbook page 4

Place value to thousands

10 tens make 1 hundred
10 hundreds make 1 thousand.
We write this as 1000.
Two thousand, three hundred and forty-nine is written as 2349.
We extend the place value table to include thousands like this:

Thousands	Hundreds	Tens	Units
2	3	4	9

1 Say these numbers aloud.
 a 2812 b 9322 c 6871
 d 3562 e 8447 f 4924

2 Write in figures:
 a five thousand, seven hundred and ninety-two   *5,792*
 b eight thousand, two hundred and seventy-five *8,275*
 c three thousand, six hundred and fifty   *3,650*
 d one thousand, nine hundred and sixty   *1,960*
 e two thousand, four hundred and eighty-nine   *2,489*
 f nine thousand, seven hundred and sixty-five   *9,765*
 g four thousand, two hundred and fifty-seven   *4,257*
 h six thousand, two hundred and eighty-three   *6,283*

3 What is the value of the red digit in each of these numbers?
 a 3612 *600*   b 9486 *6*   c 2032 *30*
 d 3009 *3,000*   e 3620 *20*   f 4915 *900*
 g 5106 *5,000*   h 9999 *90*   i 1359 *9*

you can use Workbook page 5

Working with larger numbers

We can write larger numbers using expanded form.
4765 = (4 × 1000) + (7 × 100) + (6 × 10) + (5 × 1)
     = 4000 + 700 + 60 + 5

1 Write each of these numbers in expanded form.
- a 5240
- b 1098
- c 1609
- d 3182
- e 8056
- f 7484
- g 6179
- h 2147
- i 9762

2 Write these numbers. Say each one aloud.
- a 4000 + 300 + 30 + 5
- b 8000 + 500 + 20 + 9
- c 4000 + 300 + 20 + 1
- d 5000 + 400 + 80 + 9
- e 3000 + 200 + 30
- f 2000 + 40 + 9
- g 9000 + 50
- h 7000 + 500 + 3

3 Copy these number sentences. Fill in the missing numbers.
- a 1876 = 1000 + ☐ + 70 + 6
- b 3876 = ☐ + 800 + 70 + 6
- c 4265 = 4000 + 200 + 60 + ☐
- d 9398 = ☐ + 300 + ☐ + 8
- e 3098 = ☐ + 90 + 8
- f 5008 = ☐ + 8

you can use Workbook page 6

7

Comparing and ordering numbers

Look at this number line. Do you remember how to read and use number lines like this one?

It shows the numbers from 0 to 1000.

Each small division represents one hundred.

The arrow shows the position of 300.

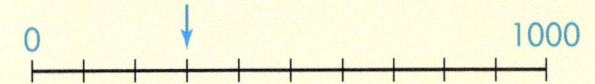

1 Write the numbers that match each letter on the number lines.

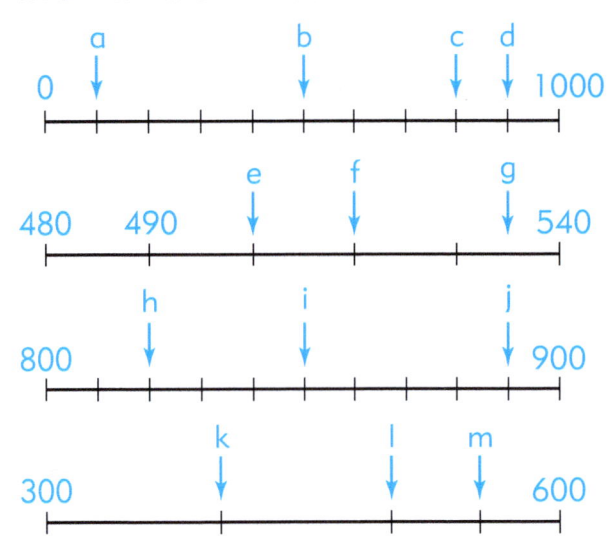

2 This 0–10 000 number line is blank.
  a Estimate the value of each number marked with an arrow.
  b Tell your partner how you decided what each number was.

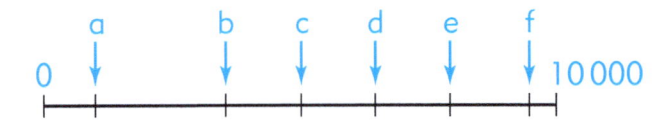

you can use Workbook page 7

Compare and order using < and >

< means 'less than',   100 < 105
> means 'greater than',   105 > 100

1 Copy and complete. Fill in < or >.

a 199 ☐ 150   b 210 ☐ 270
c 439 ☐ 498   d 760 ☐ 719
e 499 ☐ 501   f 285 ☐ 300

2 Use these number cards. Write:

a three number sentences using <

b three number sentences using >.

145   100   638   249   975   728

3 Write each set of numbers in order from smallest to greatest.

a 720, 659, 985, 460, 302

b 118, 102, 108, 111, 150

c 907, 970, 709, 790, 799, 797

4 Say whether each statement is true or false.

a 405 > 400 + 50   b 799 < 797

c 369 < (3 × 100) + (9 × 10) + (6 × 1)

d 588 < (5 × 100) + 80 + 8

you can use Workbook page 8

More ordering and comparing numbers

1 Write down the largest number in each set.

a 1326, 1632, 1623

b 4779, 4797, 4977

c 2581, 2851, 2815

d 6239, 6329, 6392

2 For each number above you wrote down, write the number that is:

a ten smaller than that number

b one hundred greater than that number

c two thousand greater than that number.

3 Rewrite each number sentence. Write digits in place of the * to make each number sentence true.

a 2460 > 2*60

b 4**9 < 4119

c 5099 < *999

d 4819 < 4*20

e 6568 < 6*6*

f *312 > 8547

4 Write a number that is in between the two numbers in each number sentence.

5 Write each set of numbers in order, from the smallest to the greatest.

a 6125 1372 5827 3150 6324

b 2895 3199 1596 2250 1826

c 5643 4291 6126 5242 4871

d 9875 8982 9645 8777 7958

e 6708 7729 5936 6599 6821

f 1986 2135 1520 2448 1992

you can use Workbook page 8

Rounding to the nearest 10

To round a number to the nearest ten, ask yourself:

Is it nearer to the next ten or the ten before?

If the units are 4 or fewer, round down.

If the units are 5 or more, round up.

27 is about 30

24 is about 20

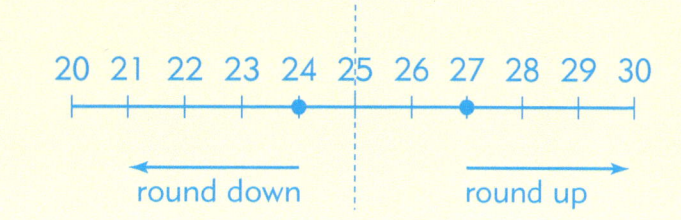

20 21 22 23 24 25 26 27 28 29 30

round down   round up

1 Round each number to the nearest ten.
You can draw number lines to help you.

a There were 148 people at the party.

b The car took 652 litres of petrol.

c Jamal made 385 cupcakes.

d 142 people bought tickets for the show.

e I have 276 books.

f 1405 people bought tickets for the show.

2 Give an example of when you might round each amount off and why.

a a price

b a height measurement

c a mass

d a time

3 Should each person round up or down? Give a reason for each answer.

a A seamstress needs 8.4 metres of fabric for some curtains. The fabric is sold in full metres.

b A caterer needs 6 kg of flour to make the bread for a party. Flour is sold in 5 kg bags.

you can use Workbook page 9

Rounding to the nearest 100

4 or fewer hundreds are rounded downwards.
5 or more hundreds are rounded upwards.

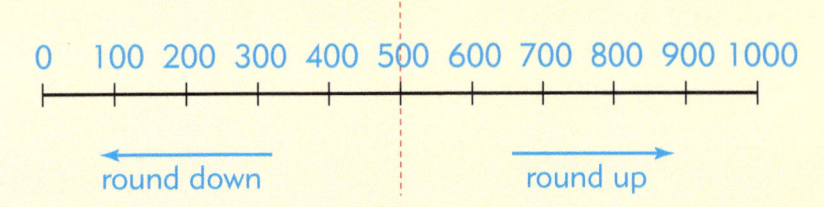

```
0  100 200 300 400 500 600 700 800 900 1000
               round down  |  round up
```

To round to the nearest hundred, look at the digit in the tens place.
If this digit is 1, 2, 3 or 4 leave the hundreds digit unchanged and write zeros as place holders for the tens and units.
For example: What is 1439 to the nearest 100?
1439   The digit in the tens place is 3, so the number rounds to 1400
If this digit is 5, 6, 7, 8 or 9, round up to the next hundred and write zeros as place holders for the tens and units.
For example, round 3689 to the nearest hundred.
The digit in the tens place is 8 so round up to the next hundred: 3700.

1 Round each number to the closest 100.

- a 4170
- b 4753
- c 4500
- d 4265
- e 4851
- f 4359
- g 4672
- h 4499
- i 4900
- j 4820
- k 4065
- l 4577

2 Round each number to the nearest hundred.
Draw a number line if it will help you.

- a 2449
- b 6921
- c 8689
- d 3162
- e 5863
- f 7500
- g 4425
- h 9267
- i 3318
- j 1282
- k 4099
- l 8989

3 These numbers have been rounded to the nearest hundred. What is the smallest, and largest, number each could have been?

- a 2300
- b 5600
- c 3100
- d 7900

you can use Workbook page 10

Revising 2D shapes

A polygon is a closed shape with three or more straight sides.

When all the sides are equal in length, and all the angles are equal in size, the polygon is regular.

Shape	Examples	Properties		
		Number of sides	Number of angles	Number of vertices
Triangle		3	3	3
Quadrilateral		4	4	4
Pentagon		5	5	5
Hexagon		6	6	6
Heptagon		7	7	7
Octagon		8	8	8

1 Count the number of sides on each shape. Write the name of the shape and say whether it is a regular polygon or not.

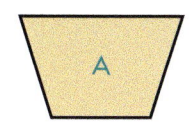

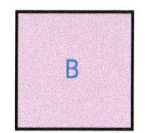

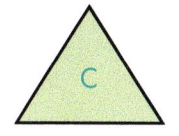

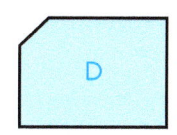

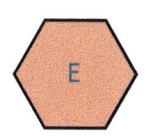

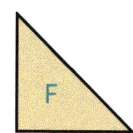

2 Draw a polygon that has:

a 6 vertices

b 4 equal sides

c a right-angle and three sides.

you can use Workbook page 11

Investigating polygons

A **pinboard** is a piece of board with nails or pegs arranged on it in a grid. You can use it to make polygons.

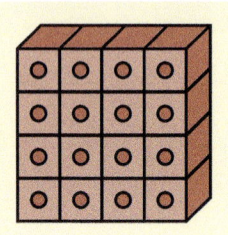

1 Use pinboards or dotted paper to make these polygons. Write their names.

a b 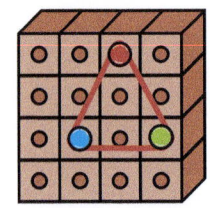 c

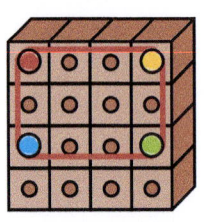

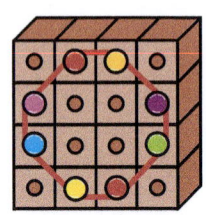

d 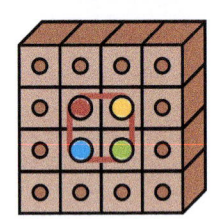 e 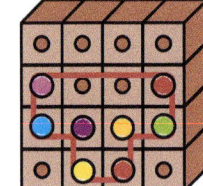 f

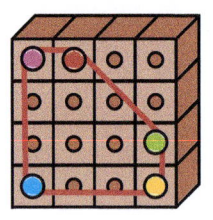

2 Find out what these road signs mean. Write what shape they are. Are they polygons?

a

b

c

d

3 Which shapes are the most common for traffic signs in your area?

you can use Workbook pages 12–13

Pinboard investigations

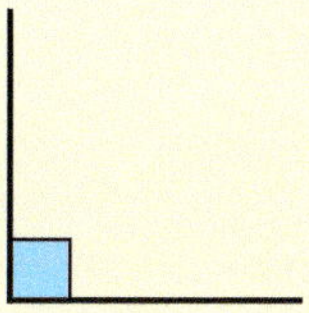

Remember: A right angle measures 90°.

The shape on this pinboard has 4 right angles and 4 sides.

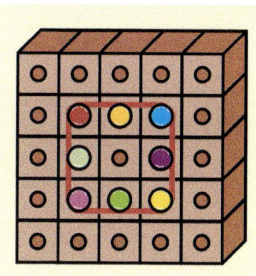

1 How many right angles do these shapes have?

a   b 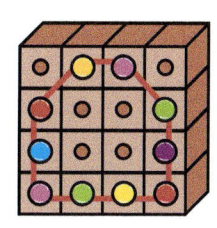   c

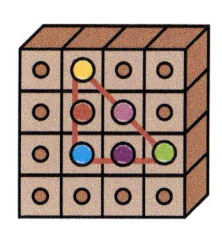

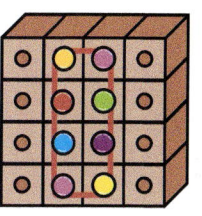

2 Use a pinboard or dotted paper to make these shapes.
   a   a quadrilateral with 1 right angle
   b   a pentagon with 3 right angles
   c   a hexagon with 2 right angles
   d   a quadrilateral with 2 right angles

3 What is the lowest number of sides a shape must have if it has 2 right angles?

4 Use your pinboard (or dotted paper) to make the following polygons. Draw each one that you make.
   a   A polygon with two right angles.
   b   A polygon with one angle that is larger than a right angle.
   c   A polygon with each angle smaller than a right angle.

More about quadrilaterals

All polygons with four sides are called quadrilaterals.
Some quadrilaterals are given special names because they have some properties that other quadrilaterals don't have.

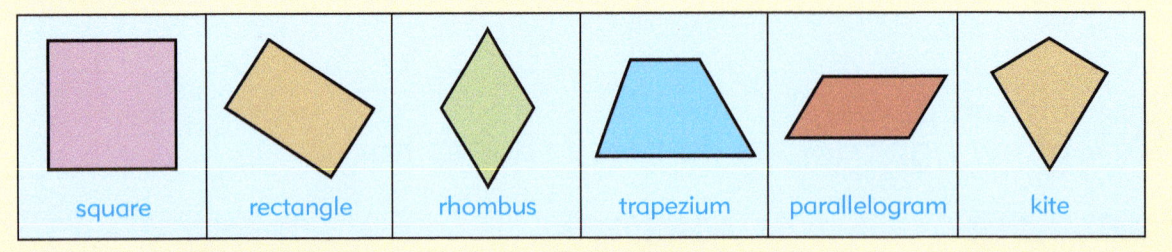

| square | rectangle | rhombus | trapezium | parallelogram | kite |

Look at the shapes.

Can you work out what special properties each of these quadrilaterals has?

1 Zahra made these shapes on dotted paper.
Write the name of each shape.

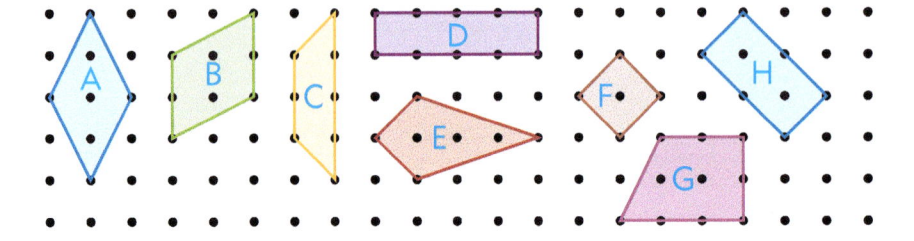

2 Read the description and then write the correct name of each shape.

a A shape with four sides.

b A regular quadrilateral.

c A quadrilateral with equal opposite sides and four right angles.

d A quadrilateral with four equal sides but no right angles.

e A quadrilateral with equal opposite sides and no right angles.

you can use Workbook page 14

Investigating rectangles

You will need:
- pinboards of different sizes
- elastic bands
- dotted paper
- used matchsticks or craftsticks.

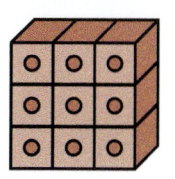

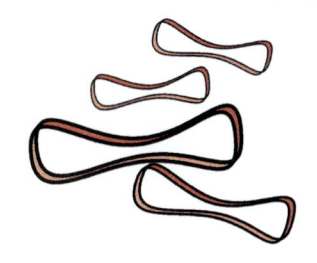

1 Use a 9-pin pinboard.

  a How many different-sized squares can you make?

  b How many other rectangles can you make?

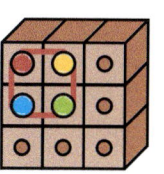

2 Now use a 16-pin pinboard.

  a How many different sized squares can you make?

  b How many other rectangles can you make?

  c Experiment with other pinboards.

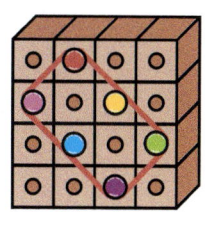

3 Make these stick patterns and solve the puzzles. Draw the new pattern.

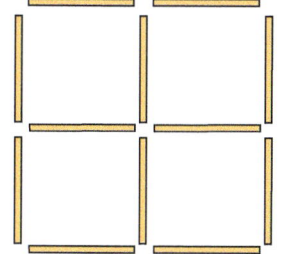

a Remove 2 sticks to leave just 2 squares.

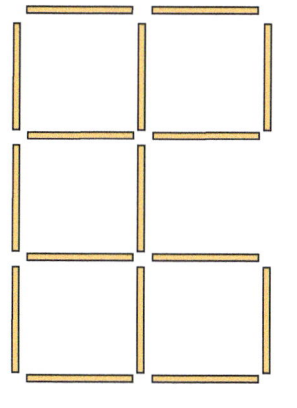

b Move 3 sticks to make 4 squares.

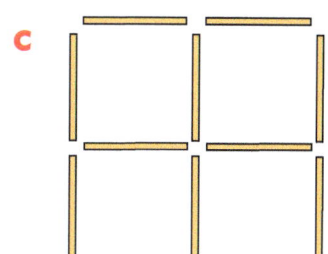

c Move 2 sticks to make 7 squares.

you can use Workbook page 15

Analogue time

A clock face with numbers and hands is known as an **analogue clock**. The minute hand makes a full turn around the clock face every hour, but the hour hand makes a full turn every 12 hours.

1 Which number on the clock face does the long hand (the minute hand) point to when it is the following times?

a half past the hour

b quarter to the hour

c quarter past the hour

d 25 minutes past the hour

e 10 minutes past the hour

f five minutes before the hour

2 Write each time in words.

a

b

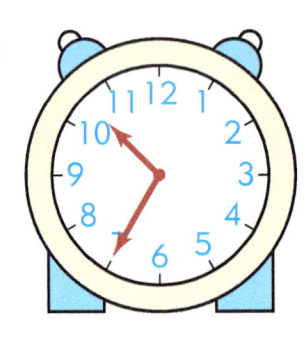

c

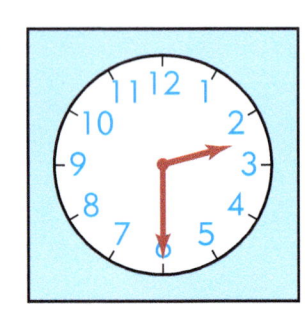

d

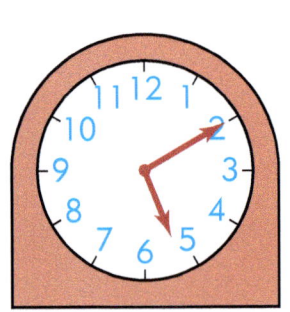

e

f

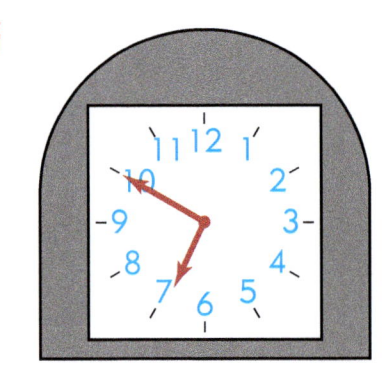

you can use Workbook page 16

Time: a.m. and p.m.

There are 12 hours before noon (a.m.) and 12 hours after noon (p.m).

9.00 AM 9 o'clock in the morning

9.00 PM 9 o'clock at night

1 For each picture, write the time in words and as a.m. or p.m time.

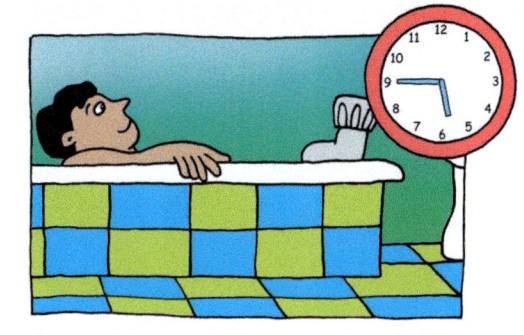

a

b

c

d

e

f

2 Write down what you do at each of these times.

19

Digital time

A **digital clock** only uses numbers to tell the time. For example:

The 'a.m.' means 'before noon'. Times from midnight to midday are a.m. times. Times from midday to midnight are p.m. times.

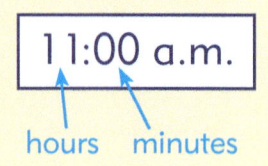

11:00 a.m.
hours   minutes

Two hours later, the time will be 1:00 p.m. or 1 o'clock in the afternoon.

1 Copy and complete the table.

Time in words	Digital time	1 hour earlier	$\frac{1}{2}$ hour later
Five o'clock in the afternoon	5:00 p.m.		05:30
Half past three in the morning		02:30	
Seven o'clock in the evening			
Twelve o'clock at night		11:00	
Quarter to one at lunchtime			01:15
Twenty to eight in the morning	7:40 a.m.		
Twenty-five past eleven at night			
Ten to five in the morning			

2 Estimate how long it would take to do the following:

a   sneeze

b   complete your secondary schooling

c   play a soccer match

d   play a test cricket game

e   get ready for school

f   travel from home to school.

20

Make your own sliding digital clock

You will need:
- strips of thin cardboard or thick paper
- a piece of cardboard
- a ruler
- scissors
- marker pens.

1 **Follow these steps.**

a Cut four strips of thin card or paper. Each strip should be about 3 cm wide. Draw blocks on the strips and label them.
- Leave one block empty at the top and bottom of each strip.
- Write the digits in the blocks as shown here.

b Make the clock face out of the thicker card.

It should be a rectangle at least 15 cm long and 10 cm wide.

Mark and make four pairs of slots that are wide enough for your strips to fit through. Look at the diagram to see how to place these.

Use the markers to draw a : between the second and third pair of slots to separate the hours and minutes.

c Thread the strips through the slots so the numbers show at the front of the card. Work in order from left to right as shown in the strips picture above.

2 **Discuss with your partner how you can use the clock you have made to show different times.**

3 **Show these times as they would appear on a digital clock.**

a twelve o'clock
b half past seven
c five to eleven
d five to nine
e quarter to six
f twenty past eight

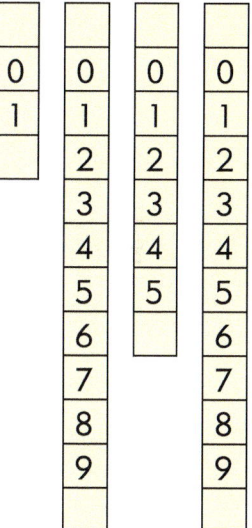

Using a calendar

This is a calendar for the month of May.

MAY						
Mon	Tues	Wed	Thurs	Fri	Sat	Sun
	1	2	3	4	5	6
7	8	9	10	11	12	13
14	15	16	17	18	19	20
21	22	23	24	25	26	27
28	29	30	31			

Thirty days has September,
April, June and November
All the rest have thirty-one
Except for February alone
It has twenty-eight days clear
And 29 in each leap year.

1 Study the calendar and answer these questions.
  a On which day of the week is the 24th?
  b What is the date of the second Tuesday?
  c Write the day and date on which the last day of this month falls.
  d What is the date the next day?
  e What is the date of the first Friday in May?

2 How many:
  a Mondays were there this month?
  b Weekends were there?

3 Sanjay wrote a test on **23 May**. He got the results a week later. When did he get his results?

4 How many days in this month fall on a date that is an odd number?

5 Look at a calendar for May of this year.
  a How is it similar to this one?
  b How is it different to this one? Why do you think it is different?

you can use Workbook page 17

22

Timetables

This is a bus timetable. It shows you eight different bus stops and the times at which the bus arrives at each stop.

Bus Timetable – Bus A232 Route 6	
Central bus station	7:15
Klimt Street	7:40
Rembrandt Square	7:55
Da Vinci Boulevard	8:03
Picasso Place	8:20
Mondrian Avenue	8:42
Escher Street	8:55
Rubens Road	9:10

1 At what time does the bus arrive at:
  a the Central bus station **b** Picasso Place **c** Rubens Road?

2 At which stop does the bus arrive at:
  a 7:40 **b** 8:42 **c** five to nine?

3 If the bus is 10 minutes late, at what time does it arrive at:
  a Klimt Street **b** Rembrandt Square **c** Escher Street?

4 If the bus is 10 minutes early, where will it be at these times?

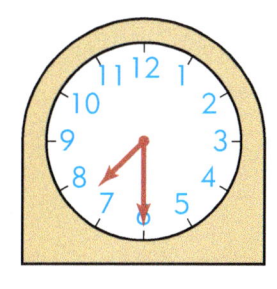

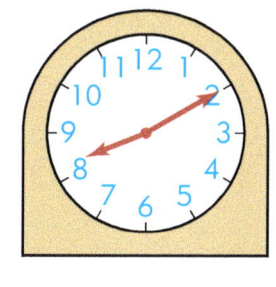

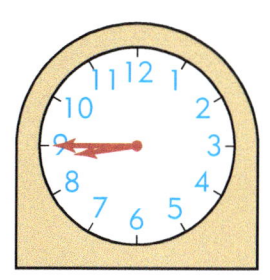

5 How long does it take the bus from:
  a the Central bus station to Klimt Street
  b Da Vinci Boulevard to Mondrian Avenue
  c the Central bus station to Rubens Road?

you can use Workbook page 18

Decimal notation

The shaded block shows one tenth or $\frac{1}{10}$
We can write this using decimal notation: 0.1
The unshaded blocks are $\frac{9}{10}$ or 0.9

100	10	1	$\frac{1}{10}$
		0 .	1

The dot is a decimal point that comes after the numeral in the units place.
Money amounts are often written using decimal notation.
$1 = 100c or 10 × 10c
Each 10c is 0.1 of $1. With money you must put an extra 0 after the decimal point like this:
$1.50 = one dollar and fifty cents = one and five tenths.

1 Write these amounts using decimal notation.
  a two dollars and two tenths of a dollar
  b six dollars and nine tenths of a dollar
  c ten dollars and one tenths of a dollar

2 Write these amounts in dollars using decimal notation.
  a 432 cents   b 809 cents   c 1250 cents
  d 1099 cents   e 2545 cents   f 2066 cents

3 How many cents are there in:
  a $4.87   b $5.99   c $12.35
  d $15.25   e $12.00   f $50.00

4 Write these amounts in order from most money to least money.
  99 cents $9.00 $0.90 $1.99

you can use Workbook page 19

Decimal place value

This picture shows a decimal abacus.

There are 2 tens → 20
3 units → 3
6 tenths → .6

It shows 23.6

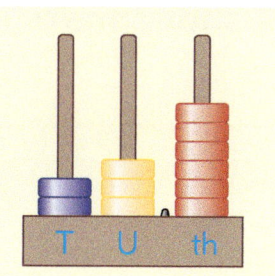

1 Write the numbers these abacus pictures show.

a   b 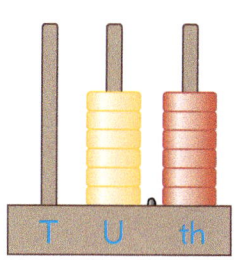   c

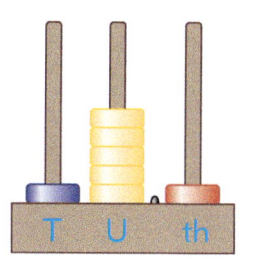

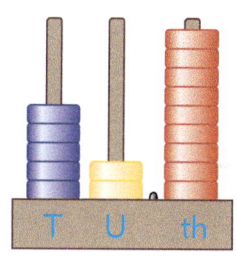

d 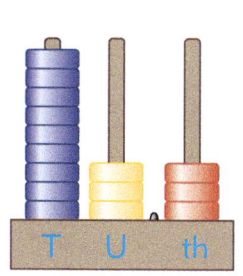   e 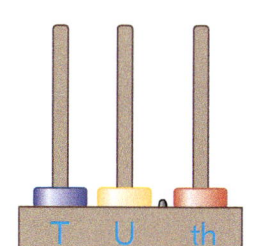   f

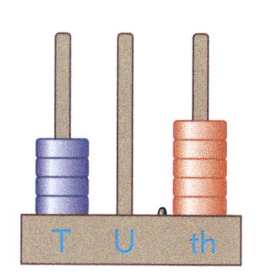

2 Draw abacus pictures to show these numbers.

a 62.4   b 5.5   c 83.0   d 20.8

3 You will need an abacus or a drawing of an abacus and some beads.

a Using only three beads, how many different decimal numbers can you make? Here is one possibility.

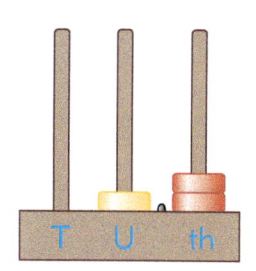

b Try again using four beads, five beads, and so on.

25

Counting in tenths

This number line shows mass in kilograms.

0   0.5   1   1.5   2   2.5   3   3.5   4   4.5   5 kg

Each small division on the number line shows $\frac{1}{10}$ of a kilogram or 0.1 kilograms.

1 Use the number line.

  a Count in tenths of a kilogram from 0.7 kg to 1.6 kg. Write all the weights.

  b Count back in tenths of a kilogram from 2 kg to 1.1 kg. Write all the weights.

2 Now do these in the same way.

  a 2.5 kg to 4.3 kg   b 0.6 kg to 1.9 kg   c 3.8 kg to 2.6 kg

  d 4.0 kg to 2.8 kg   e 2.9 kg to 3.3 kg   f 2.1 kg to 0.9 kg

3 Order these weights from heaviest to lightest.

  a 6.8 kg, 5.9 kg, 7.1 kg

  b 15.3 kg, 20.5 kg, 9.7 kg, 19.7 kg

  c 86.8 kg, 87.9 kg, 86.5 kg, 87.1 kg

  d 31.4 kg, 59.0 kg, 39.1 kg, 45.2 kg, 51.1 kg

4 Order these weights from lightest to heaviest.

  a 9.8 kg, 4.3 kg, 11.1 kg, 7.4 kg

  b 15.3 kg, 9.4 kg, 12.8 kg, 10.0 kg

  c 43.3 kg, 39.1 kg, 36.8 kg, 41.0 kg, 36.2 kg

  d 19.9 kg, 18.7 kg, 19.1 kg, 18.3 kg, 19.6 kg

you can use Workbook page 20

Decimals and fractions

100 square

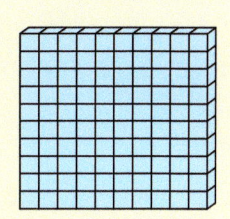

1 whole
one
1.0

If 100 squares make 1 whole, then the 10 square strip is $\frac{1}{10}$ of the whole.

We write tenths as decimals like this:
$\frac{1}{10} = 0.1$

50 blocks out of 100 is the same as five 10s. $\frac{50}{100} = \frac{5}{10}$

We write $\frac{5}{10}$ as a decimal like this: 0.5

But 50 out of 100 is also half of the square, so $\frac{5}{10} = \frac{1}{2}$

That means that $\frac{5}{10} = \frac{1}{2} = 0.5$

10 long

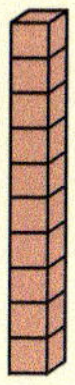

$\frac{1}{10} = 0.1$

1 Write each amount as a decimal fraction.

a 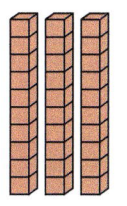
b 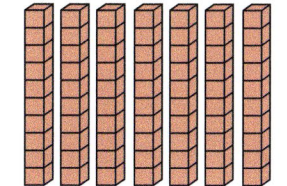
c 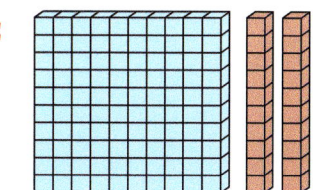
d 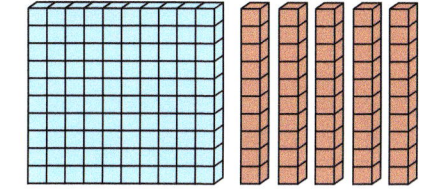
e

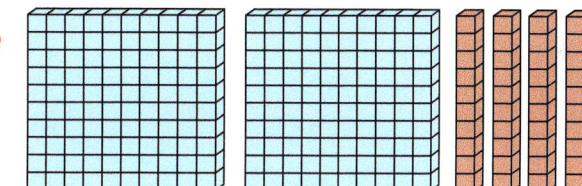

2 How many tenths are shown on each abacus?

a   T U th
b   T U th
c   T U th

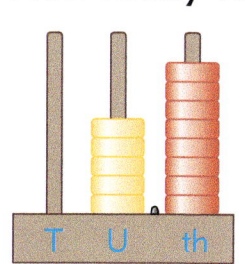

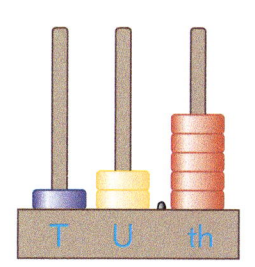

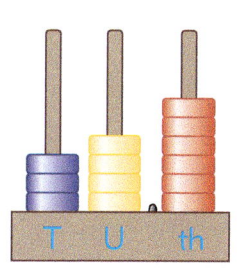

3 Write the amount shown by the coloured arrows on the number line as a decimal fraction.

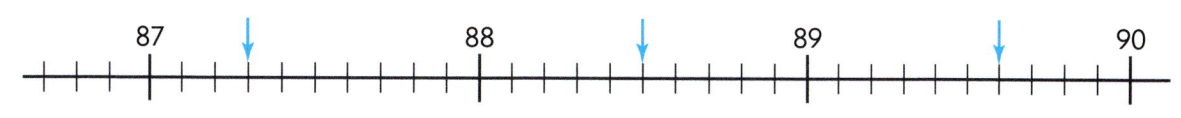

87   88   89   90

you can use Workbook page 21

27

More decimals

100c = $1   150c = $1.50   105c = $1.05

1 Write these amounts in dollars ($).
- a 175c
- b 225c
- c 309c
- d 580c
- e 1260c
- f 2505c
- g 1995c
- h 4999c
- i 7004c

2 Write these amounts in cents (c).
- a $0.50
- b $0.05
- c $5.55
- d $4.70
- e $15.25
- f $29.99

3 Amina measured her height. She wrote it in two ways:

Name	cm	m
Amina	136	1.36

Measure your height and the heights of four friends. Make a chart. Show your heights in two ways.

4 Use a square marked in hundredths. Create an interesting design by colouring in. Colour 0.24 of the square in one colour and 0.76 in another.

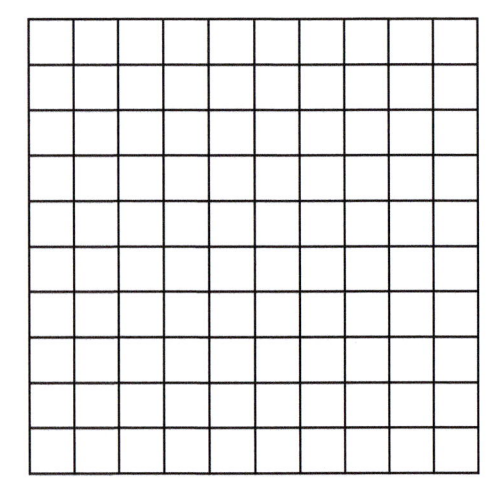

Comparing decimals

0.12 kg > 0.09 kg

0.12 kg is heavier than 0.09 kg
0.12 > 0.09

0.22 kg < 0.28 kg

If the tenths are the same, look at the hundredths.
0.22 kg is lighter than 0.28 kg
0.22 < 0.28

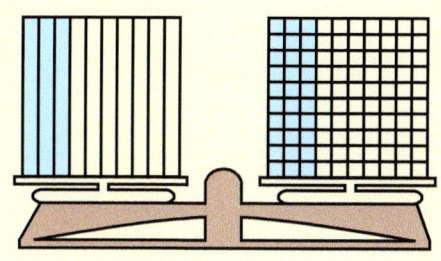

0.3 kg = 0.30 kg

Sometimes weights are equal.
0.3 = 0.30

1 Each square represents one kilogram. Write the amount shaded as a decimal. Then use >, < or = to make true statements.

a

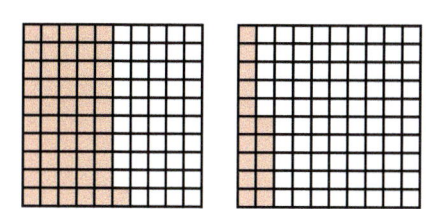

b

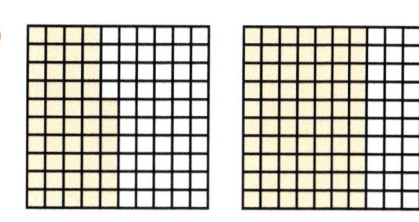

c

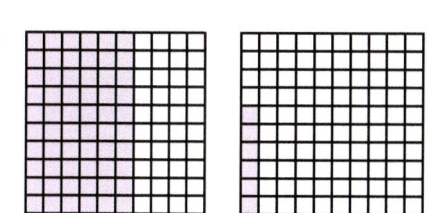

d

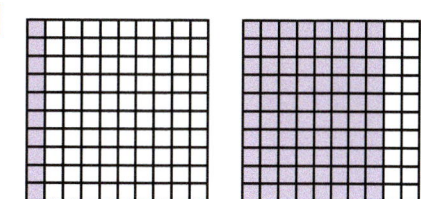

2 Write >, < or = to make true statements.

a 0.26 m ___ 0.52 m
c 0.64 m ___ 0.46 m
b 0.13 m ___ 0.13 m
d 0.07 m ___ 0.7 m

you can use Workbook page 22

Gymnastics scores

There are four main events in this gymnastics competition.

Here are the scores of five competitors.

	Vault	Bars	Floor	Beam
Francesca	8.50	7.65	8.00	7.15
Barbara	7.85	8.35	8.25	7.40
Mino	8.25	7.80	8.50	8.20
Denise	8.10	8.55	9.10	8.80
Olga	7.95	8.25	8.75	8.50

1 Use the information on the chart to help you answer these questions.

  a What is the highest score for each competitor?

  b Write each person's scores in order from greatest to lowest.

2 Who do you think won the gold, silver and bronze medals? Explain how you decided.

Metric units of length

1 cm = 10 millimetres (mm)
1 cm = 1 centimetre (cm)
100 cm = 1 metre (1 m)
1000 m = 1 kilometre (1 km)

1 Use a ruler. Draw:

a a flower 4 cm high
b a stick 7 cm long.

2 The units are missing from the measurements in these sentences. Work with a partner to decide what the missing units are in each example.

a At 11 years old, Sarah was 142 ____ tall.

b After driving for 68 ____, Marcel reached the next town.

c When Asif had sharpened his pencil, the point was 5 ____ long.

d Anita needed another 10 ____ strip of card to finish her model.

e My fingernail is 9 ____ wide.

f Abdul has a new mobile phone that is 8 ____ thick.

g Nita has just finished a 100 ____ run.

h Khalid and Sara walk about 225 ____ to school.

you can use Workbook page 23

31

Reading and writing lengths

Julie measured her height.
She was 1 m and 19 cm tall.

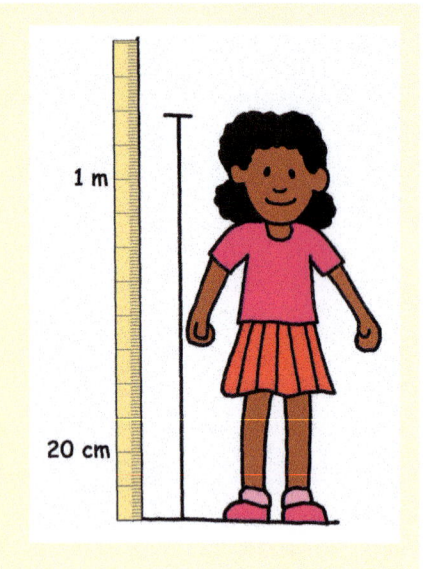

Julie's height can be written in three ways:

1 m 19 cm or   119 cm or   1.19 m

Look at this ruler. The centimetres are numbered and shown with a longer line. The millimetres are shown, but they are not numbered. You have to count them to work out how long the green line is.

The green line is 2 cm and 9 mm long.
This is the same as 2.9 cm or 29 mm.

1 Measure your own height. Write it in three ways.

2 Write each of these measurements in three different ways.

The car is 250 cm long.

The table is 1 metre and 47 cm long.

3 What is the length of each of these green lines? Write each length in three different ways.

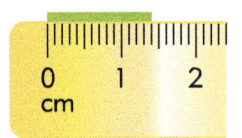

you can use Workbook page 24

Litres and millilitres

The **capacity** of a container tells us how much it can hold. The standard unit of capacity is the litre.

1 litre (1 l) = 1000 millilitres (1000 ml)

Measuring spoons and cups have exact capacities.

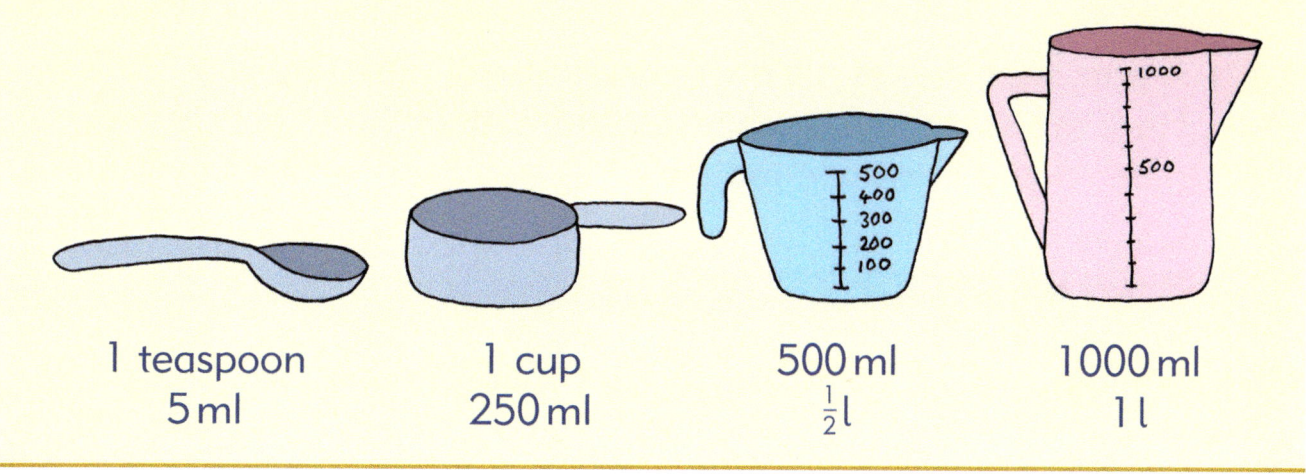

| 1 teaspoon | 1 cup | 500 ml | 1000 ml |
| 5 ml | 250 ml | $\frac{1}{2}$ l | 1 l |

1 Collect some containers from yoghurt, juice, milk and other liquids.

a Arrange your containers in order from the one with the smallest capacity to the one with the biggest capacity. Try to estimate the capacity of each container.

b Use measuring jugs and spoons to measure the exact capacity of each container.

2 You can check the answers to these questions using real measuring spoons and cups.

a How many teaspoons in 1 cup?

b How many cups in 1 litre?

c How many cups in 2.5 litres?

3 How many litres do you need to fill:

a 8 cups   b 9 cups   c 11 cups   d 22 cups?

Mixing amounts

Remember: There are 1000 millilitres in a litre.

The short way of writing millilitres is ml

The short way of writing litres is l

1 A professor makes his own experimental mixtures. Here are some of his mixtures. Write down how much liquid each one makes.

a Mixture A
250 ml beetroot juice
175 ml bath water
50 ml custard

b Mixture B
350 ml pond water
50 ml beetroot juice
150 ml custard
200 ml ink

c Mixture C
150 ml beetroot juice
250 ml custard
220 ml bath water
125 ml ink

d Mixture D
550 ml pond water
250 ml beetroot juice
350 ml custard
425 ml bath water
375 ml ink

2 The professor makes each mixture. How much of each bottle of ingredients does he have left?

3 Use the leftover ingredients to invent your own mixture. How much liquid do you make?

Using liquids

1 Work out the answer to each of these number problems.

a Jennifer has to take 10 ml of her medicine 4 times a day. She has a 250 ml bottle of medicine. How long will it last?

b A light aircraft burns 75 l of fuel every hour. How many litres of fuel are used on a 3-hour flight?

c A leaky tap drips 1.25 l of water every half hour. How much is wasted in 5 hours?

d A small car uses 1.75 l of petrol every 20 kilometres. How much does it use to travel 200 kilometres?

2 How can you use these buckets to measure 4 litres of water?

7 l   5 l   3 l

Which is best?

These instruments are all used for weighing.

1 kg 575 g can be written as 1.575 kg

1 Write down what you would use to weigh the following objects. Also write down whether you would measure in kg, g or kg and g.

2 Estimate the mass of each item and then measure and record it.

a

b

c

d

e

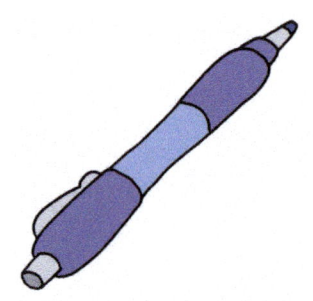

f

g

you can use Workbook page 25

36

Reading measuring scales

Look at this scale.
The kilograms are numbered.
There is one division between each numbered kilogram.
This division shows $\frac{1}{2}$ kg or 500 g.
These potatoes weigh 2 kg and 500 g.
You can write the weight in three ways:

2 kg 500 g   or   2500 g   or   2.5 kg

Remember: 1 kg = 1000 g so $\frac{1}{2}$ kg = 500 g

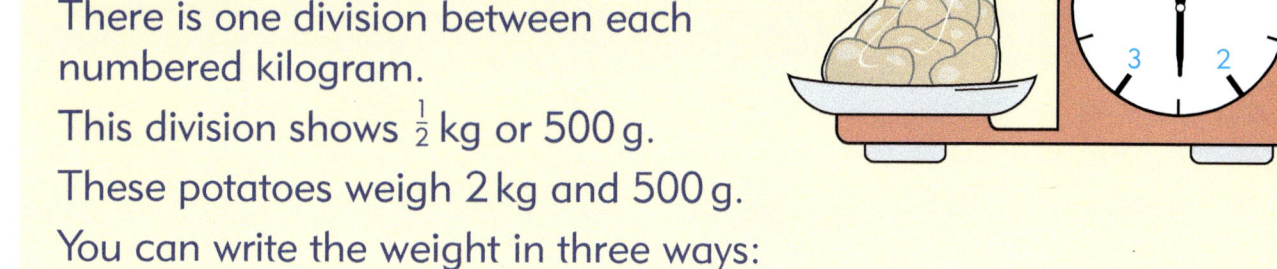

1 What weights do these scales show?

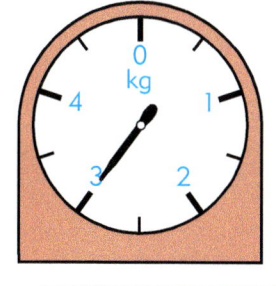

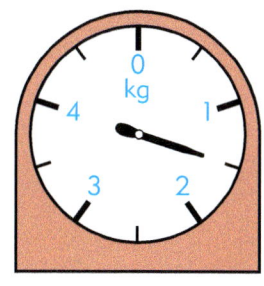

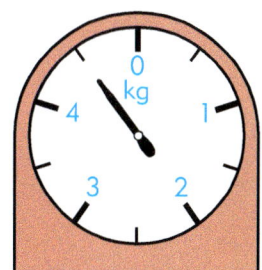

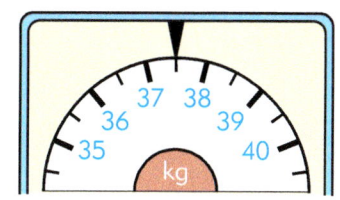

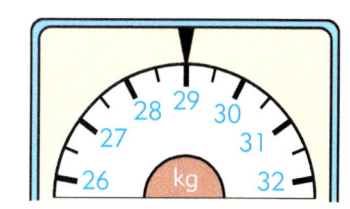

2 Work with a partner. Look at these scales.

a What weight does each small marking represent?

b Write down the weights marked A to H.

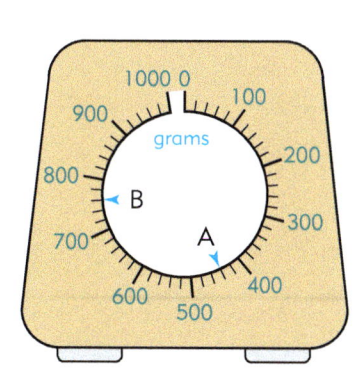

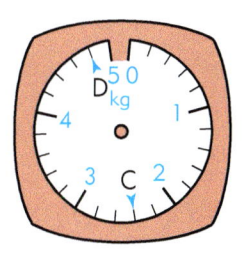

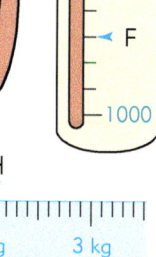

you can use Workbook pages 26–27

37

Counting on and back

Do you remember how to count on and back in ones, tens and hundreds from any number?
Look at the digits that change in each of these counting patterns.

Count on in ones: 1234   1235   1236
Count on in tens: 1234   1244   1254
Count on in hundreds: 1234   1334   1434
Count back in ones: 3688   3687   3686
Count back in tens: 3688   3678   3668
Count back in hundreds: 3688   3588   3488

Now that you are working with bigger numbers, you can also count on and back in thousands. Look at these examples:

Count on in thousands from 3700 until you pass 8000.

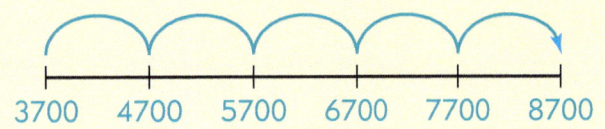

3700   4700   5700   6700   7700   8700

Count back in thousands from 9500 until you pass 5000.

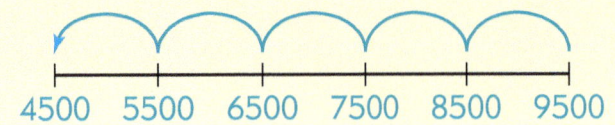

4500   5500   6500   7500   8500   9500

1 **Count on:**

a 9 in ones from 1288
b 90 in tens from 3000
c 900 in hundreds from 1200
d 5000 in thousands from 1455

2 **Count back:**

a 8 in ones from 1000
b 80 in tens from 2890
c 800 in hundreds from 4500
d 8000 in thousands from 9100

3 Write the number that is 1000 less than each number and the number that is 1000 more than each number.

a 4567   b 8045   c 5055

4 Starting at 1245, how many hundreds do you need to count to pass 2000?

you can use Workbook page 28

Adding or subtracting groups of 10s, 100s and 1000s

Read through these examples.

What is 2529 + 30?

30 is 3 tens, so I will count on in tens.

2529   2539   2549   2559

What is 9600 − 4000?

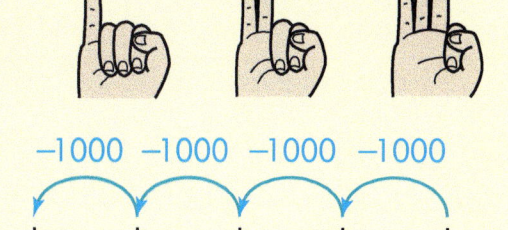

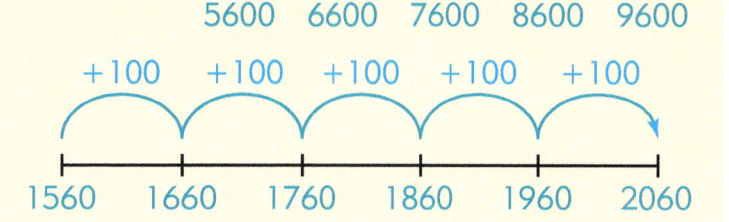

5600 6600 7600 8600 9600

Add 1560 + 500

1560 1660 1760 1860 1960 2060

1 Add.

a 1234 + 40   b 3044 + 500   c 400 + 3560
d 20 + 3456   e 50 + 1280   f 900 + 3200
g 400 + 899   h 1345 + 8000   i 700 + 1200

2 Subtract.

a 8600 − 400   b 987 − 60   c 7350 − 200
d 9866 − 5000   e 8990 − 4000   f 6543 − 400
g 2010 − 20   h 4345 − 400   i 5645 − 70

3 Complete the number sentences.

a 4300 − ☐ = 4100   b 2080 − ☐ = 2030
c ☐ − 3000 = 4680   d 580 + ☐ = 1080
e 2450 + ☐ = 2950   f 3020 + ☐ = 3100

4 Write down one number you could add to get these results:

a 1235 → 1735   b 3448 → 3498   c 3250 → 8250
d 4567 → 4867   e 9200 → 10 000   f 4860 → 5160

Making 100

You already know all the pairs of numbers that make 10:
1 + 9 = 10   2 + 8 = 10   3 + 7 = 10   4 + 6 = 10   5 + 5 = 10

We can use these facts to find pairs of multiples of ten that make 100.

10 + 90 = 100   20 + 80 = 100   30 + 70 = 100
40 + 60 = 100   50 + 50 = 100

We can also use the facts we already know to find any pair of numbers that make 100.

For example:

36 + ☐ = 100

You can think like this:
I know 36 + 70 will give me 106
70 − 6 = 64

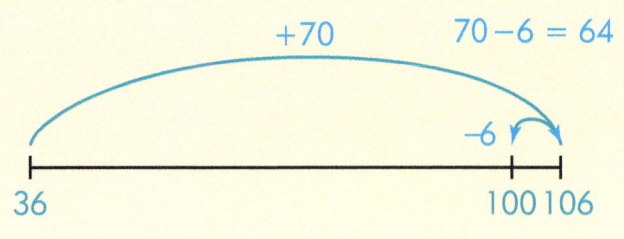

You can also think like this:
100 − 36
= 70 − 6 = 64

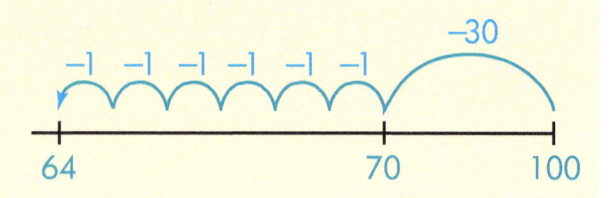

1 Find the missing numbers in each number sentence.

a 72 + ☐ = 100
b 33 + ☐ = 100
c 45 + ☐ = 100
d ☐ + 19 = 100
e ☐ + 64 = 100
f ☐ + 61 = 100

2 The weight of each pile of rocks is given. Work out how many kilograms you would need to add to each pile to make 100 kilograms.

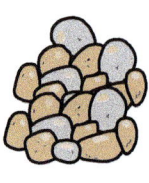

57 kg   39 kg   87 kg   13 kg

you can use Workbook page 29

Pairs of numbers that make 1000

We can use the addition facts we know to quickly find pairs of numbers with a total of 1000.

1 + 9 = 10	10 + 90 = 100	100 + 900 = 1000
2 + 8 = 10	20 + 80 = 100	200 + 800 = 1000
3 + 7 = 10	30 + 70 = 100	300 + 700 = 1000
4 + 6 = 10	40 + 60 = 100	400 + 600 = 1000
5 + 5 = 10	50 + 50 = 100	500 + 500 = 1000

Can you see the pattern in these sums?
What about numbers that end in 50?

150 + ☐ = 1000
150 + 50 = 200
200 + 800 = 1000
So, 150 + 850 = 1000

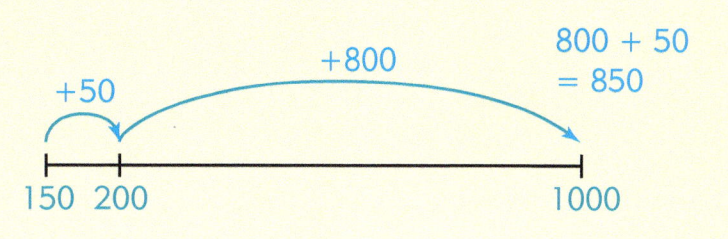

+50   +800   800 + 50 = 850

150 200   1000

1 Write the matching pairs that make 1000.

450   950   400   350   750   150   700

700   550   100   650   850   300   200

2 There are 1000 millilitres in one litre. How much juice will be needed to fill each of these 1 litre containers?

a 150 ml
b 400 ml
c 650 ml
d 850 ml

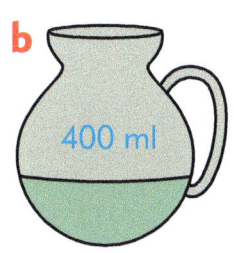

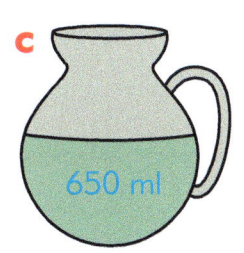

Revising addition facts

Answer these questions about the picture. Try to work them out in your head. Then write the number sentences in your exercise book.

1. How many children are skipping?

2. How many children are playing games with a ball?

3. If the tennis players joined in skipping, how many children would be skipping?

4. How many children are there altogether?

5. Find four groups that can be put together to make 30.

6. Find five groups that can be put together to make 20.

7. List all the groups which could be put together to make 11.

8. How many groups of six could all the children make? Would any be left over?

you can use Workbook page 30

Adding numbers by making 10 or 20

3 + 4 + 6 = 13   3 + 6 + 4 = 13   6 + 3 + 4 = 13

You can add numbers in any order.

When you have to add three or four small numbers you can add them in pairs.

If you find pairs that add to 10 or 20, it makes addition easier.

6 + 5 + 4   Add 6 + 4 first   13 + 8 + 7   Add 13 + 7 first
10   20
10 + 5 = 15   20 + 8 = 28

4 + 9 + 7 + 1   Add 9 + 1 first
10
10 + 7 = 17
17 + 4 = 21   17 ⚪⚪ count on if you need to.

1 Add these numbers. Try to find pairs that make 10 or 20 and add them first.

a 7 + 11 + 3   b 9 + 7 + 1   c 5 + 8 + 5
d 8 + 6 + 2   e 7 + 5 + 3   f 6 + 13 + 4
g 2 + 11 + 8   h 8 + 7 + 12   i 9 + 6 + 11
j 7 + 5 + 13   k 15 + 6 + 5   l 14 + 7 + 6

2 Add these numbers.

a 4 + 7 + 3 + 9   b 6 + 6 + 8 + 2
c 6 + 8 + 14 + 3   d 9 + 5 + 11 + 3
e 12 + 7 + 8 + 3   f 13 + 9 + 7 + 1

43

Adding multiples of 10

You know that 2 + 3 = 5. So 20 + 30 = 50

2 tens + 3 tens = 5 tens which is 50

Like you did with smaller numbers, you can group pairs of tens to make it easier to add.

3 + 6 + 7   30 + 60 + 70   3 tens + 7 tens = 10 tens which is 100
= 10 + 6   = 100 + 60
= 16   = 160

1 One long rod is made of 10 cubes. How many cubes are on the desk?

a

b

c

d

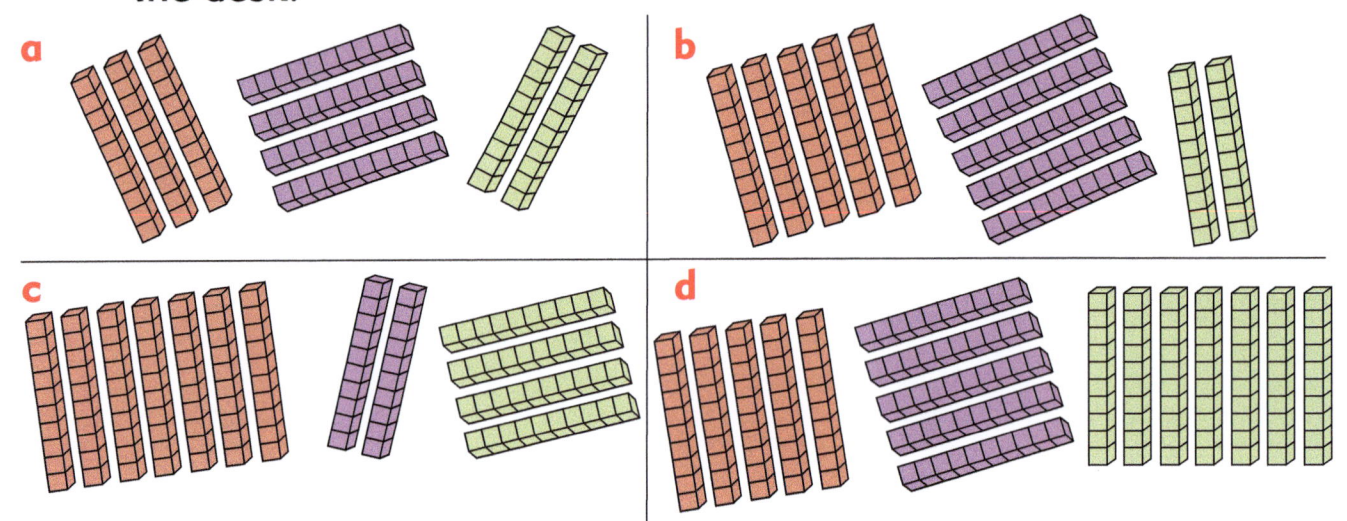

2 Add these numbers as quickly as you can. When you have finished, tell your partner how you worked out the answers.

a 20 + 20 + 20   b 30 + 20 + 50   c 40 + 80 + 60
d 30 + 60 + 40   e 90 + 40 + 10   f 40 + 80 + 20

3 Try to do these sums mentally.

a 30 + 40 + 20   b 40 + 30 + 30   c 40 + 30 + 30
d 40 + 10 + 50   e 50 + 10 + 20   f 20 + 30 + 30
g 40 + 30 + 40   h 30 + 30 + 30

Symmetry

Do you remember that a line of symmetry is a line which divides a shape in half so that each half is the mirror image of the other?

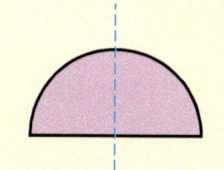

The dashed line is the line of symmetry.

If you fold the shape along the line of symmetry, the two halves will fit exactly onto each other.

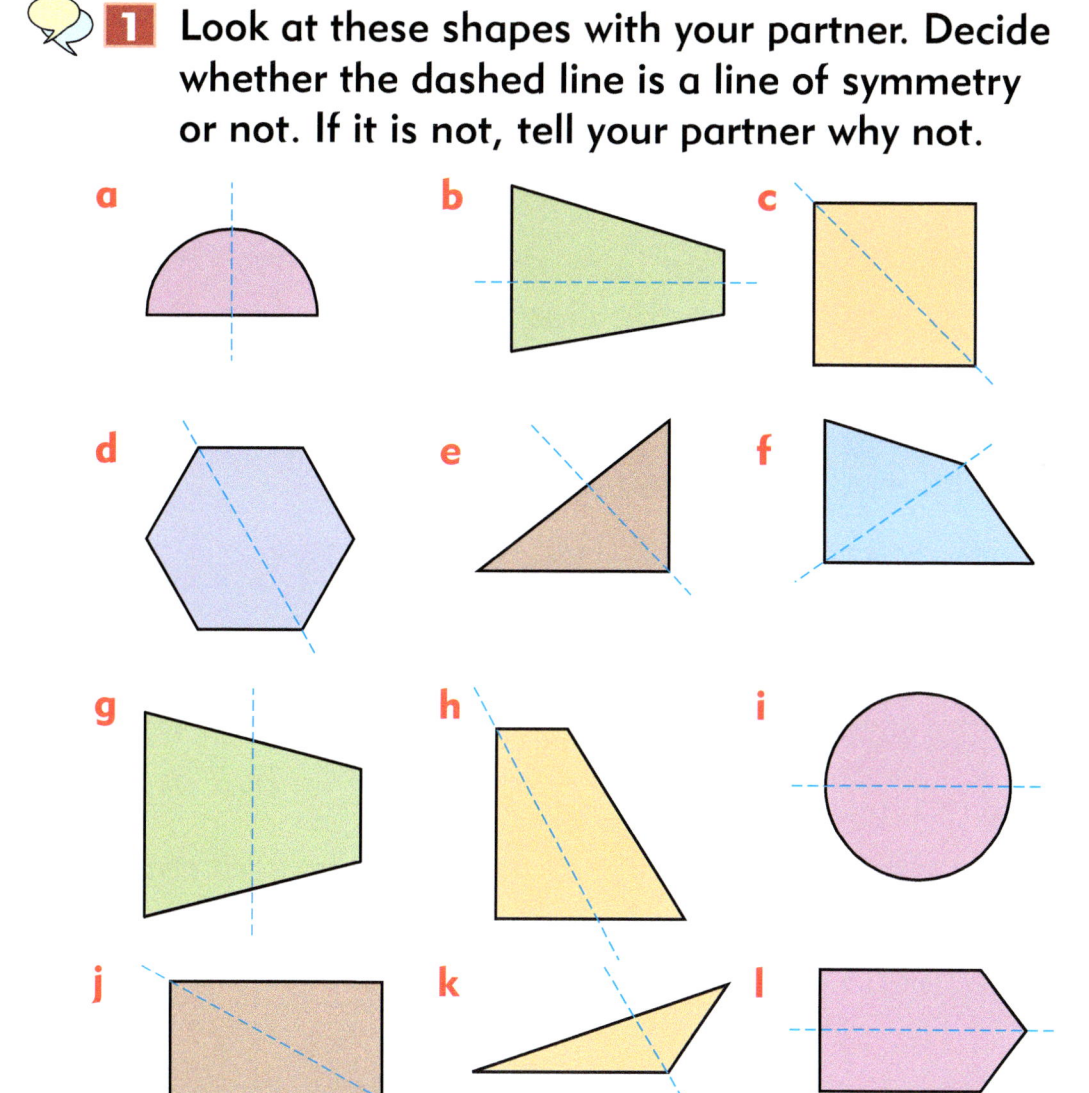

1 Look at these shapes with your partner. Decide whether the dashed line is a line of symmetry or not. If it is not, tell your partner why not.

a   b   c

d   e   f

g   h   i

j   k   l

Symmetry in polygons

1 **Here are five regular polygons.**

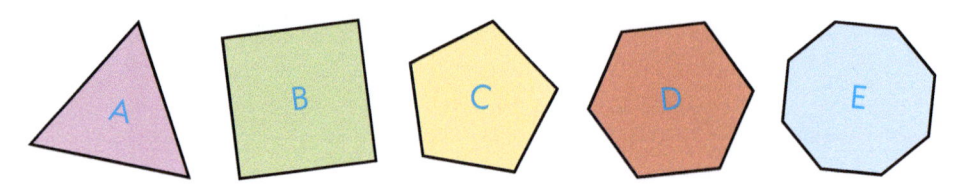

a Write the name of each shape.
b Say whether it is symmetrical or not.
c Draw each shape in your book. Draw the lines of symmetry that you can use to fold each shape and get two identical halves. Use a different colour for each line of symmetry.
d What is the connection between the number of sides of these shapes and the number of lines of symmetry they have?

2 **A square has four sides and four lines of symmetry. You can see these in the diagram. Does this mean that all quadrilaterals have four lines of symmetry?**

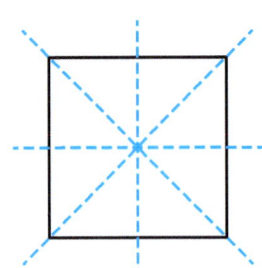

a Draw these quadrilaterals on dotted paper.

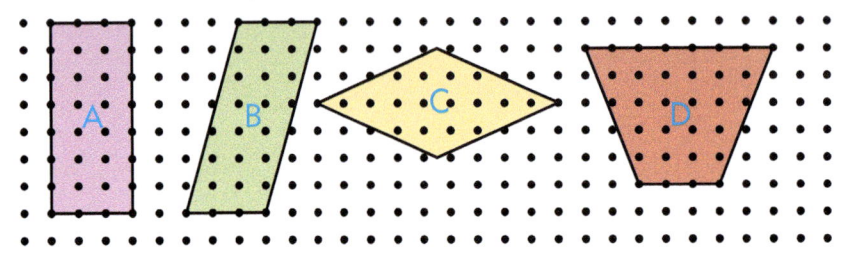

b Write the name of each shape below it.
c Draw the lines of symmetry in each shape. Use a different colour for each line.
d Why don't these quadrilaterals have four lines of symmetry?

46

Symmetry around us

There are many examples of symmetry in daily life.

These buildings have been designed to be symmetrical. Can you find the lines of symmetry?

1 Find two examples of symmetrical buildings in your community. Sketch or photograph them and draw the lines of symmetry.

2 This is a rangoli pattern from India. These are symmetrical, traditional folk-art designs which are still made today.

   a What shapes can you find in the pattern?

   b How many lines of symmetry can you find?

3 These pictures show halves of symmetrical shapes. The dashed line is the line of symmetry. Work out how many sides the completed shapes would have and name them.

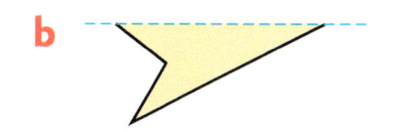

a   b   c

you can use Workbook page 31

Organising data

Do you remember how to use tally marks?

A tally is a line that you make each time you count an item. If you count three items you will make three tallies like this: ///

Each time you count five you draw the tally across the previous 4 marks like this: ////

This makes it easier to add up all the tallies by counting in fives.

A teacher asked students in his class which two pieces of fruit they would each like to take with them on the class picnic. To help him buy the fruit, he made this tally chart.

Fruit	Tally
Apples	//// //// //// ////
Pears	//// //// ///
Oranges	//// //// ////
Bananas	//// ////
Peaches	//// ////
Mangoes	//// //

1 How many of each of these fruits would the teacher need to buy?

a apples   b pears   c oranges
d bananas   e peaches   f mangoes

2 Remember each student in the class was allowed to choose two fruits. How many students are there in the class altogether?

3 Imagine your class is going on a picnic.
  a List six fruits that you can buy locally.
  b Ask 10 students to choose two fruits that they would like to take on a class picnic.
  c Make a tally chart to record the results.

Frequency tables

The **frequency** of a value is the number of times it occurs. We can use a frequency table like the one below to show how many times each value occurs. The frequency is the total of the tallies.

1 A class of 20 students got the following marks out of 10 for an assignment:

7 6 6 7 5 8 10 9 9 7

7 7 6 5 7 10 8 7 6 7

a Copy this frequency table.

Mark	Tally	Frequency
5		
6		
7		
8		
9		
10		

b Complete the second column of the table by tallying how frequently each mark occurs.

c Write the total in the frequency column.

2 Find out the ages of 20 students in your class (in years). Draw a simple frequency table to show your results.

3 Do a survey of what time your friends go to bed on school nights, rounded off to the nearest half-hour. Draw a simple frequency table to show your results.

49

A typical day

Maria drew this bar chart of how she spent one school day.

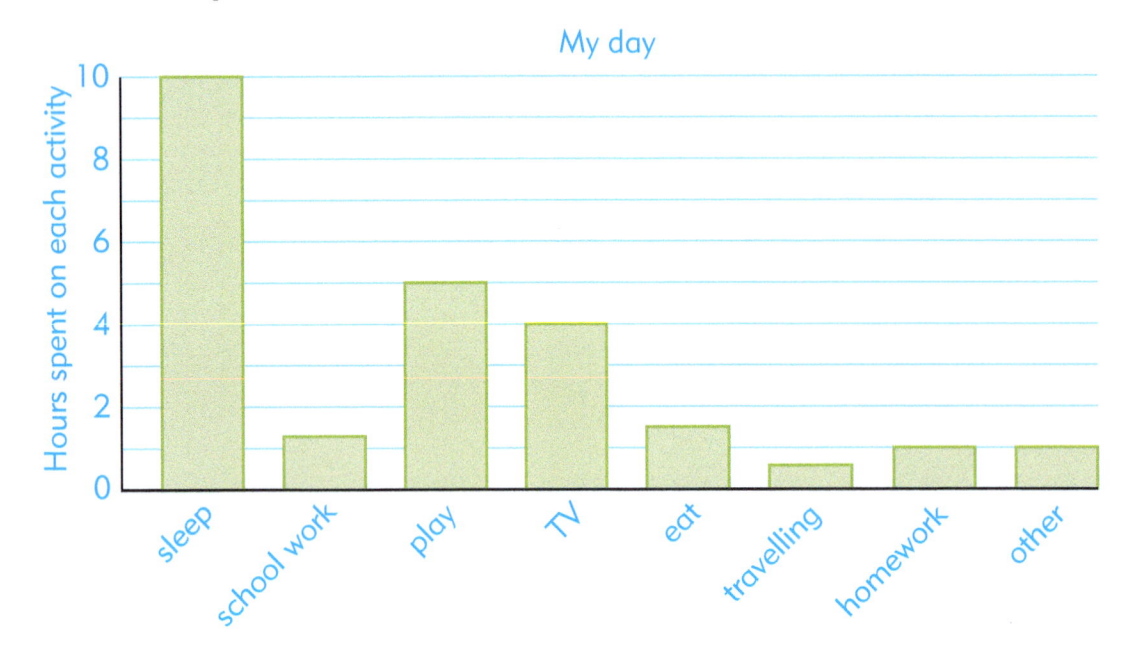

My day

(Bar chart: Hours spent on each activity — sleep: 10, school work: 1, play: 5, TV: 4, eat: 1.5, travelling: 0.5, homework: 1, other: 1)

1 Answer the questions.

a How long did she spend eating?

b How long did she choose to watch TV?

c How much time did her homework and school work take up?

d What did she spend the longest time doing?

e What did she spend the shortest time doing?

2 Draw your own bar chart showing how long you spend on different activities during the day.

Who uses the shop?

A newsagent wanted to find out how many people used his shop in the school holidays.

He counted the people who came into his shop between 12:30 p.m. and 1:30 p.m. Here is a bar chart showing the results of his survey.

People visiting the shop

1 How many of each of these groups were there?

a girls
b boys
c women
d men
e babies

2 How many people visited the shop altogether?

3 How many more boys visited the shop than:

a girls
b men
c babies?

4 Do you think everyone who visited the shop was a customer? Give a reason for your answer.

5 How do you think the bar chart would look if the newsagent collected the information during the school term?

More bar charts

The coach kept a tally of how many goals the strikers scored in soccer matches in one season. He asked James and Spike to each draw a bar chart to compare the results.

Sipho												
Jabu													
James																				
Rashid	||||| ||||| ||||| ||||| |||||	
Solomon																					
Spike														

This is the bar chart James drew.

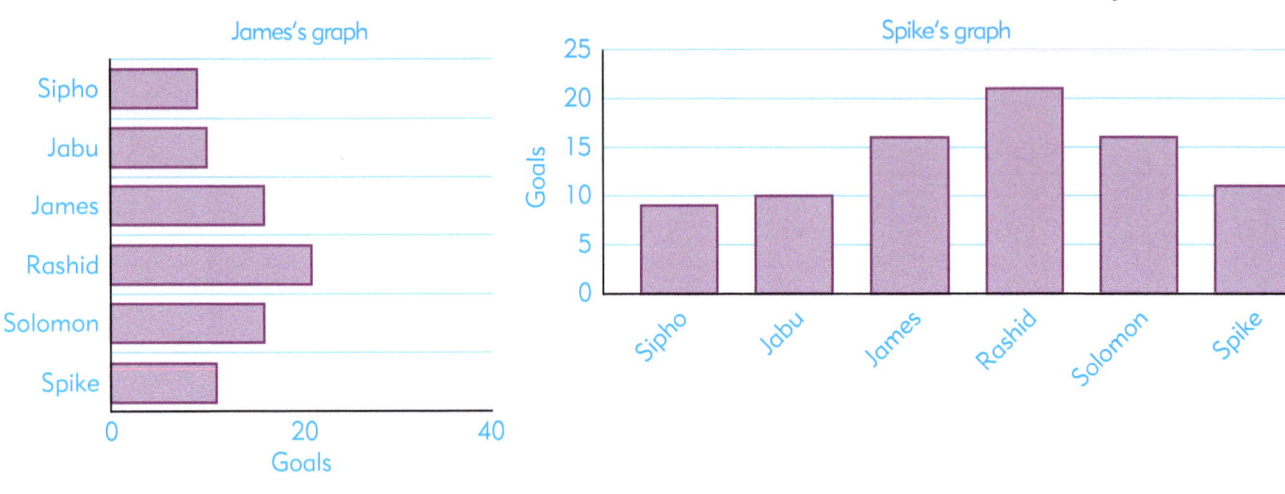

This is the bar chart Spike drew

1 Compare the two bar charts.
  a Are both bar charts correct? Give reasons for your answer.
  b Which bar chart do you think is clearer? Why?
  c What makes the bar charts look different?

2 This table shows how much money a taxi driver spent on petrol each month for five months.

September	October	November	December	January
$170	$130	$90	$110	$90

Draw a bar chart to show this information. Use a scale of 1 cm for $20 on the vertical axis.

3 How would the bar chart look if the scale was:
  a 1 cm for $50? b 1 cm for $5?

you can use Workbook page 32

52

Reading a pictogram

	Ice creams sold in one week
Vanilla	🍦🍦🍦🍦🍦🍦🍦🍦
Chocolate	🍦🍦🍦🍦🍦🍦🍦🍦🍦🍦🍦
Strawberry	🍦🍦🍦🍦🍦🍦
Mint	🍦🍦🍦🍦🍦🍦

🍦 = 10 ice creams

Look at the pictogram and answer the questions.

1 Which was the most popular flavour?

2 Which two flavours were equally popular?

3 How many more vanilla cones were sold than strawberry?

4 How many more chocolate cones were sold than mint?

5 Do you think the graph shows the exact number of ice cream cones sold?
Give a reason for your answer.

6 How could a graph like this help someone running an ice cream shop?

53

More pictograms

1 This pictogram shows the number of ships arriving at a small harbour on different days of the week.

Number of ships arriving	
Monday	🚢 🚢 🚢 🚢
Tuesday	🚢 🚢
Wednesday	🚢 🚢 🚢
Thursday	🚢 🚢
Friday	🚢 🚢 🚢 🚢
Saturday	🚢 🚢 🚢 🚢 🚢 🚢
Sunday	🚢

🚢 = 5 ships

a On which day did most ships arrive?
b On which two days did the same number of ships arrive?
c How many ships arrived on Monday?
d How many more ships arrived on Saturday than on Friday?
e How many ships arrived altogether this week?

2 Complete this frequency table using information from the pictogram above.

Days of the week	Frequency
Monday – Thursday	
Friday	
Saturday	
Sunday	

3 This pictogram shows the number of different animals that a group of tourists saw on a game drive.

Animals we saw on our game drive	
Lions	◖
Antelope	⊕ ⊕ ⊕ ⊕ ◖
Giraffes	⊕ ◖
Elephants	⊕ ◕

⊕ = 20 animals

a Draw the symbols you would use to show:
   • 10 animals
   • 5 animals
   • 15 animals.
b How many different types of animals did they see?
c How many antelope did the group see?
d Which animal did they see five of?
e How many more giraffes did they see than lions?

you can use Workbook page 33

54

Drawing pictograms

Use the grids on Workbook page 33 to complete these activities.

1 Sarah collected this information.

Hair colour	Tally	Frequency
Black	𝍤 𝍤 I	11
Dark brown	𝍤 III	8
Light brown	𝍤 II	7
Other	IIII	4

a Complete the pictogram to show this information.
b Which hair colour was most common? How can you tell this from the pictogram?
c How many students did not have black hair?
d How many more students had black hair than light brown hair?
e Why does the pictogram need to have a key?

2 Kyrill drew this bar chart to show how many books he and his friends read during the summer holidays.

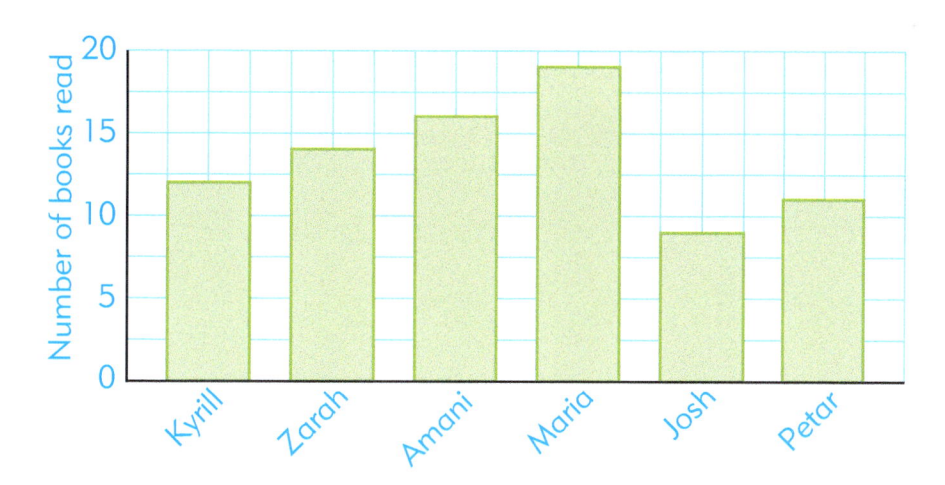

a Using a symbol of ☐ = 4 books, draw a pictogram to show the same data.
b Compare the bar chart and the pictogram. Discuss the advantages of each type of graph.

you can use Workbook page 33

Revising repeated addition and subtraction

What is 6 lots of 5?
You can find this by adding six lots of 5: 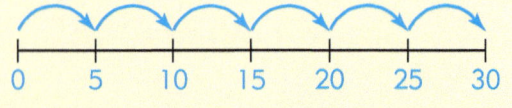
You can also find 6 lots of 5 by multiplying.
6 × 5 = 30
Multiplication is the same as repeated addition.
How many groups of 4 can you make from 24?
You can find this by subtracting groups of 4 till you get to 0.
So you can make 6 groups of 4. 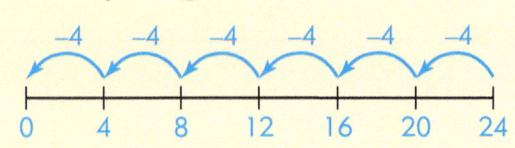
You can also find this by dividing.
24 ÷ 4 = 6

This picture shows ten groups of 4. Use it to help you answer the questions.

1 What is:
  a 2 groups of 4
  b 4 groups of 4
  c 8 groups of 4
  d 5 groups of 4

2 How many groups of 4 can you make from:
  a 12
  b 28
  c 36

3 Calculate:
  a 4 × 5
  b 4 × 6
  c 8 × 4
  d 9 × 4
  e 8 × 2
  f 8 × 3

4 How many groups of 8 can you make from:
  a 16
  b 24
  c 40

Revising the 2×, 3×, 5× and 10× tables

You should already know the multiplication facts for the 2×, 3×, 5× and 10× tables. Use these lists to revise them if you need to.

2 × 1 = 2	3 × 1 = 3	5 × 1 = 5	10 × 1 = 10
2 × 2 = 4	3 × 2 = 6	5 × 2 = 10	10 × 2 = 20
2 × 3 = 6	3 × 3 = 9	5 × 3 = 15	10 × 3 = 30
2 × 4 = 8	3 × 4 = 12	5 × 4 = 20	10 × 4 = 40
2 × 5 = 10	3 × 5 = 15	5 × 5 = 25	10 × 5 = 50
2 × 6 = 12	3 × 6 = 18	5 × 6 = 30	10 × 6 = 60
2 × 7 = 14	3 × 7 = 21	5 × 7 = 35	10 × 7 = 70
2 × 8 = 16	3 × 8 = 24	5 × 8 = 40	10 × 8 = 80
2 × 9 = 18	3 × 9 = 27	5 × 9 = 45	10 × 9 = 90
2 × 10 = 20	3 × 10 = 30	5 × 10 = 50	10 × 10 = 100

The answer you get when you multiply two numbers is called the product of those numbers.

1 Write down the answers. Try not to look at the tables.

a 5 × 4
b 2 × 9
c 3 × 8
d 10 × 9
e 5 × 6
f 2 × 10
g 5 × 2
h 5 × 8
i 3 × 6
j 2 × 3
k 3 × 3
l 5 × 3
m 5 × 5
n 3 × 3
o 10 × 10

2 What is the product of:

a 2 and 6
b 3 and 5
c 10 and 4
d 3 and 7
e 5 and 4
f 10 and 8

3 Write the missing numbers.

a 2 × ☐ = 20
b 3 × ☐ = 27
c 5 × ☐ = 45
d 3 × ☐ = 30
e 8 × ☐ = 40
f 7 × ☐ = 70

The 4× table

You already know some of the multiplication facts for 4.

2 × 4 = 8   so 4 × 2 = 8
3 × 4 = 12   so 4 × 3 = 12
5 × 4 = 20   so 4 × 5 = 20
10 × 4 = 40   so 4 × 10 = 40

You can count in fours to find the other multiplication facts for the 4× table.

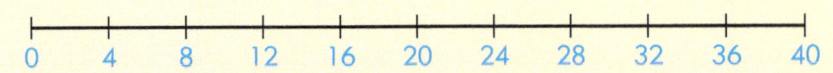

```
0   4   8   12  16  20  24  28  32  36  40
```

1 Copy this table list into your book.

1 × 4 =   2 × 4 =
3 × 4 =   4 × 4 =
5 × 4 =   6 × 4 =
7 × 4 =   8 × 4 =
9 × 4 =   10 × 4 =

2 Work out the missing answers. You can use the pictures on page 56 to help you.

3 Learn the 4× table by heart.

4 Solve these problems using multiplication facts for 4.

a Sandra buys 6 books that cost $4 each. How much does she pay?
b A packet contains 10 stickers. Nick buys ten packets. How many stickers is that?
c Four children have 9 counters each. How many is that altogether?
d Four times a number is 36. What is the number?
e A fish bowl holds 4 litres of water. How many bowls can you fill with 28 litres of water?
f If 10 × 4 is 40, what is 11 × 4? How do you know this?

you can use Workbook page 34

The 6× and 9× tables

You also need to learn the 6× and 9× tables.

The 6× table

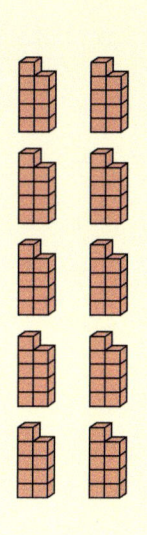

6 × 1 = 6
6 × 2 = 12
6 × 3 = 18
6 × 4 = 24
6 × 5 = 30
6 × 6 = 36
6 × 7 = 42
6 × 8 = 48
6 × 9 = 56
6 × 10 = 60

The 9× table

9 × 1 = 9
9 × 2 = 18
9 × 3 = 27
9 × 4 = 36
9 × 5 = 45
9 × 6 = 54
9 × 7 = 63
9 × 8 = 72
9 × 9 = 81
9 × 10 = 90

1 It is very important to learn your tables because these multiplication facts help you multiply bigger numbers. They also help you divide quickly.

a Make a set of small cards by cutting up some stiff paper or card.

b Work with a partner. Test each other to see which multiplication facts you know.

c If you don't know a fact, write it on one side of the card.

$$4 \times 8$$

d Look up, or work out, the correct answer. Write it on the back of the card.

e Keep the cards with you and keep testing yourself until you know all the facts.

you can use Workbook page 35

How many ways?

The arrangement of dots below is called a **multiplication array**. Look at these facts about the number 24.

6 rows of 4 = 24
4 rows of 6 = 24
6 × 4 = 24   4 × 6 = 24
4 + 4 + 4 + 4 + 4 + 4 = 24
6 + 6 + 6 + 6 = 24
24 ÷ 4 = 6   24 ÷ 6 = 4

1 Write all the facts about these arrays.

a, b, c, d

2 Use buttons, beads or stones to show these numbers as arrays.

a 32   b 45   c 48   d 36

Multiplication grids

You can use grids drawn on squared paper to help you find the answers to multiplication problems.

Look at these two grids.

This grid shows 2 × 4
You can count the blocks to find the product. 2 × 4 = 8.
You can also count blocks and use the grid to find 2 × 1; 2 × 2 and 2 × 3

This grid shows 3 × 8
You can count the blocks to find the product. 3 × 8 = 24.
You can also use this grid to find other products. For example 3 × 3 or 6 × 2.

1 Here is a blank 10 × 10 grid of squares. Use it to find each product.

a 5 × 7
b 8 × 8
c 3 × 9
d 6 × 8
e 4 × 10
f 8 × 9
g 7 × 7
h 9 × 5
i 10 × 6

2 Explain how you could use the grid above to help you work out:

a 11 × 4

b 12 × 5

Code breakers

×	6	9	7	4
5	O	W	T	V
8	E	L	M	A
3	U	S	D	I
10	N	C	P	K

This grid shows a code.
To use it, you multiply two numbers to find a letter.

5 × 4 = 20

20 is the code for V

1 Use the grid above to break this code.

45   48   72   3 × 4   40   8 × 6

56   18   9 × 8   35   12   70   72   4 × 3   90   32   5 × 7   12   30   60

8 × 4   6 × 10   3 × 7   21   12   20   12   3 × 9   3 × 4   5 × 6   60

Use the grid to make your own coded message.

Ask a friend to decode it.

You will need: 10 counters and a stopwatch.

2 Table strips game

Each player makes a table strip like the one on page 35 of the Workbook, with all the numbers filled in.

Decide on how many numbers on the strip will be covered. The first player places counters on the strip. The other player has ten seconds to work out which numbers are covered. This player removes the counters from those numbers which are correct.

The score is the number of counters collected. Swap over. Now the other player must try to collect counters. The player with the most counters is the winner.

62

Division facts

You can use the multiplication facts you know to find the division facts for those numbers.

For example, you know that 6 × 8 = 48
So, 48 ÷ 6 = 8
And 48 ÷ 8 = 6
To find a division fact such as 48 ÷ 6 ask yourself 'what times 6 gives me 48?'

This can be written as ☐ × 6 = 48
or 6 × ☐ = 48

1 Use the multiplication facts to find the missing numbers in the division sentences.

a 4 × 6 = 24   24 ÷ 4 = ☐   24 ÷ 6 = ☐
b 5 × 7 = 35   35 ÷ 5 = ☐   35 ÷ 7 = ☐
c 6 × 7 = 42   42 ÷ 6 = ☐   42 ÷ 7 = ☐
d 9 × 4 = 36   36 ÷ 9 = ☐   36 ÷ 4 = ☐
e 10 × 8 = 80   80 ÷ 10 = ☐   80 ÷ 8 = ☐

2 Complete these number sentences.

a ☐ × 5 = 40   b ☐ × 3 = 27   c ☐ × 4 = 16
d ☐ × 2 = 18   e ☐ × 6 = 42   f ☐ × 7 = 21
g 5 × ☐ = 25   h 8 × ☐ = 56   i 6 × ☐ = 54
j 4 × ☐ = 28   k 6 × ☐ = 48   l 10 × ☐ = 100

3 Write the answers to these divisions.

a 30 ÷ 6 = ☐   b 35 ÷ 7 = ☐   c 63 ÷ 9 = ☐
d 24 ÷ 6 = ☐   e 90 ÷ 9 = ☐   f 30 ÷ 5 = ☐
g 21 ÷ 3 = ☐   h 45 ÷ 5 = ☐   i 16 ÷ 4 = ☐
j 27 ÷ 3 = ☐   k 56 ÷ 8 = ☐   l 81 ÷ 9 = ☐

you can use Workbook page 37

Revising 3D shapes

3D shapes are the solid shapes we see around us.

- 3D shapes have flat or curved surfaces that we call faces.
- The faces meet at an edge.
- The edges meet at a corner or point called a vertex.

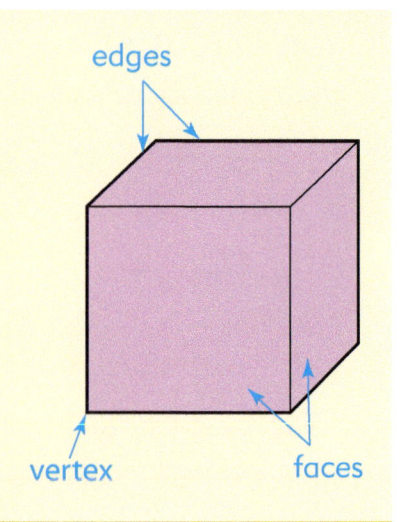

Copy and complete this table for the 3D shapes shown below. If you cannot work out the answers, find objects that are the same shape and use them to count the parts.

Shape	A	B	C	D	E	F	G	H
Number of faces								
Number of edges								
Number of vertices								

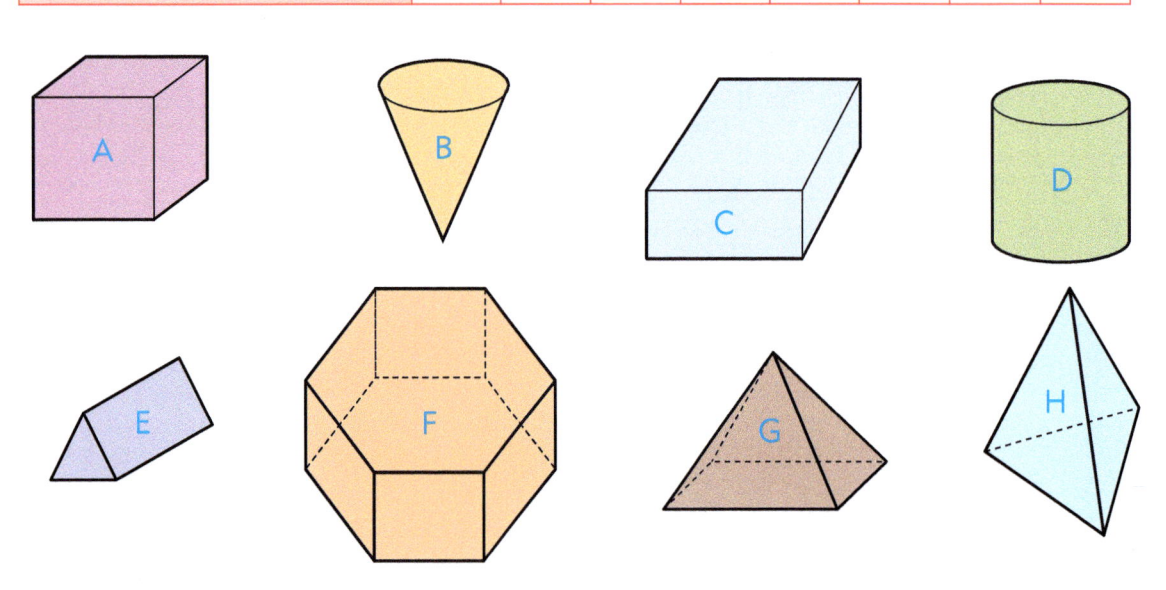

Naming 3D shapes

Cubes and cuboids

These are box shapes. A cube is a square box shape. The faces are all squares of the same size.

A cuboid is a rectangular box shape. The faces are all rectangles.

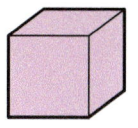

cube

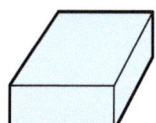

cuboid

Prisms

A prism is a shape with ends that are the same shape and size. The other faces are rectangles. The shape of the end face gives the prism its name.

This is triangular prism.

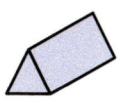

This is a hexagonal prism.

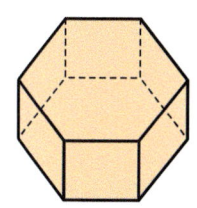

Pyramids

Pyramids have one flat base. The other faces are triangles that meet at one vertex.

The shape of the base is used to name the pyramid.

This is a square-based pyramid.

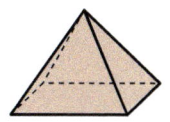

This pyramid has a triangular base which is the same shape and size as the faces. It can be called a triangular pyramid or a tetrahedron.

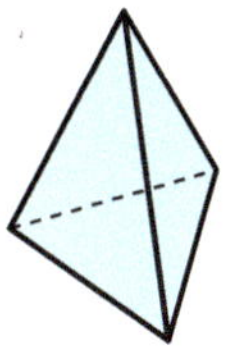

Cones and cylinders

A cylinder has two flat circular faces and one curved surface that flattens out to form a rectangle.

A cone has one flat circular face and a curved surface that flattens out into part of a circle.

This is a cone.

This is a cylinder.

1 Name these solids.

a 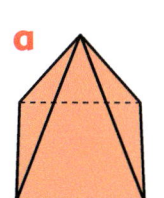   b 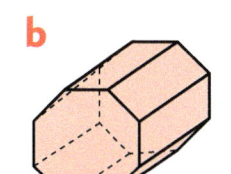   c 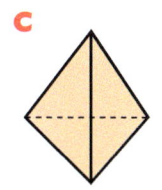   d 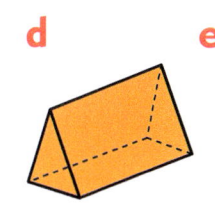   e 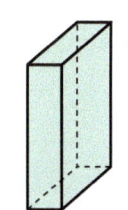   f 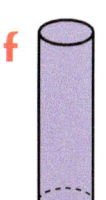   g

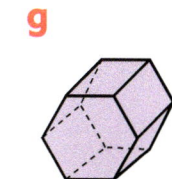

Solids and their nets

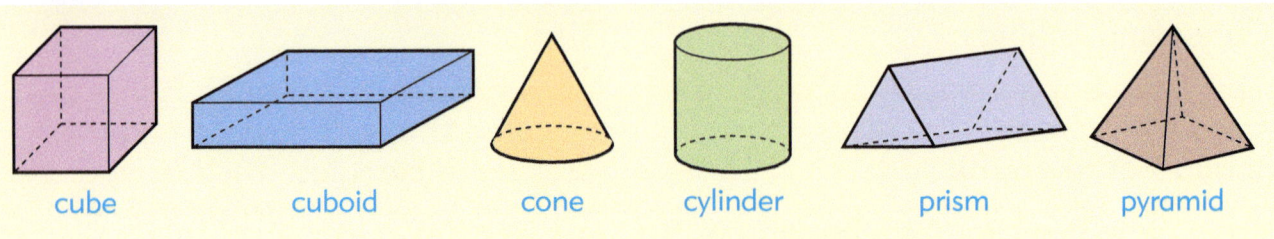

cube   cuboid   cone   cylinder   prism   pyramid

A **net** is a flat shape that you can fold to enclose a solid shape.

1 Say what solid each net would enclose when you fold it up. Copy the shapes and make your own net for each shape.

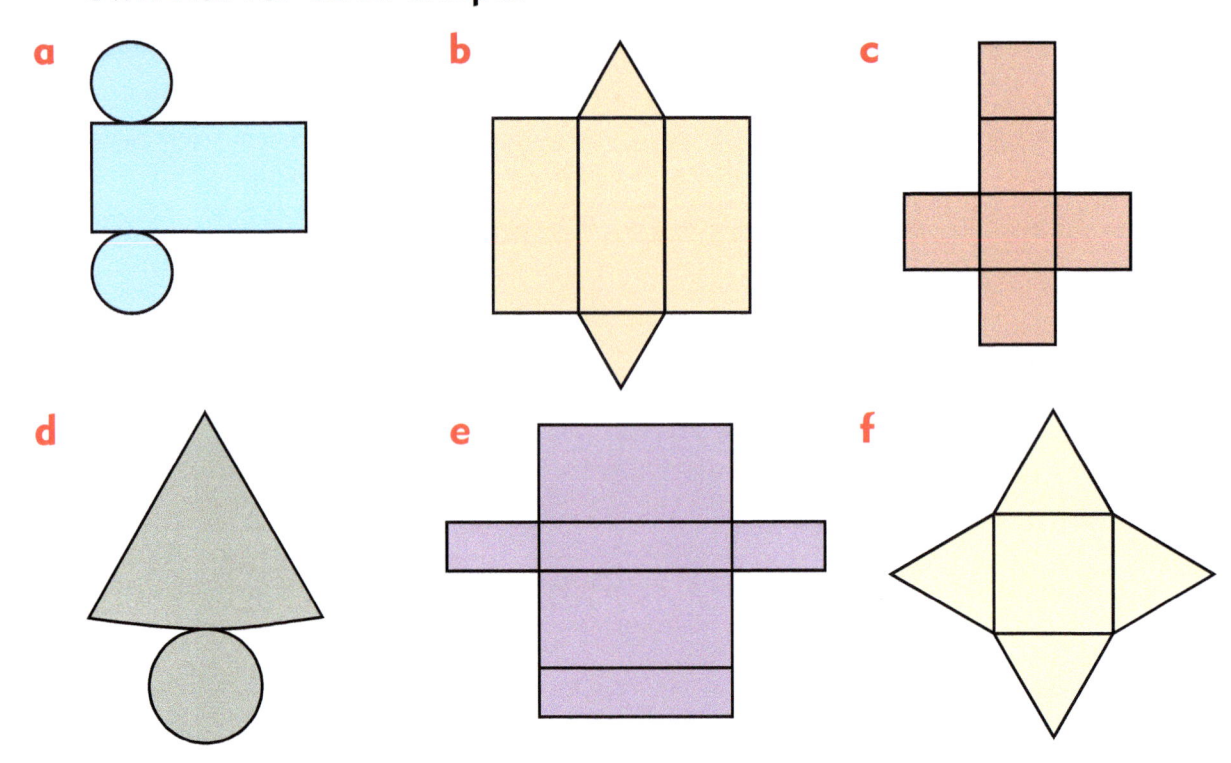

a   b   c

d   e   f

2 Use squared paper. Choose one of the solids above and draw two different ways of making its net.

you can use Workbook page 38

Is it a cuboid?

A cuboid has:
- 6 rectangular faces
- right-angled vertices
- opposite faces that are the same size and shape.

A **cube** is a special sort of cuboid because all its faces are identical.

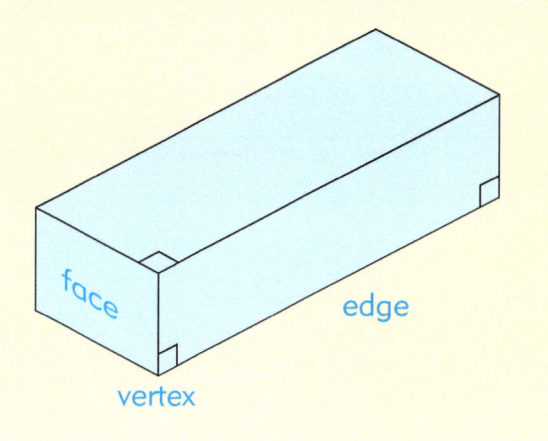

1 Say which of these shapes will make a cube or cuboid.

a

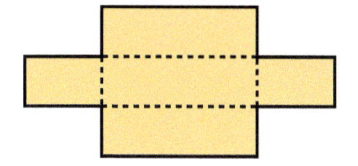

b

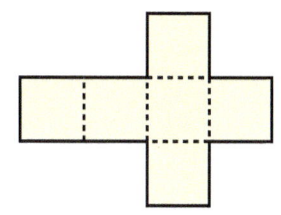

c

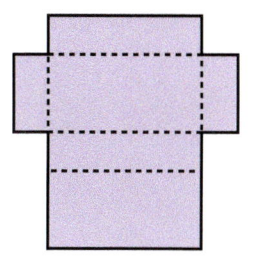

d

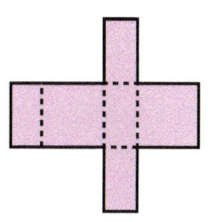

e

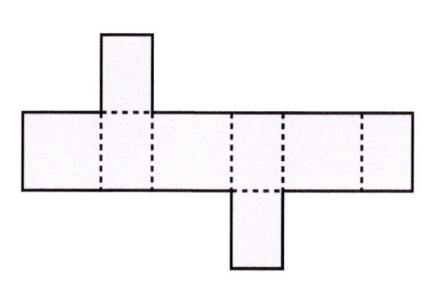

f

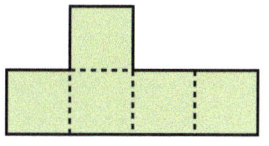

2 Draw each net, cut it out and try to make the shapes.

67

Is it a pyramid?

The base of a pyramid is a polygon. All the other faces are triangles that meet at a point. We name each pyramid according to the shape of its base.

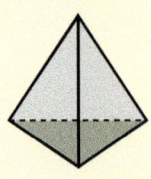

triangular pyramid 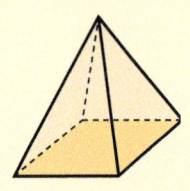   square-based pyramid 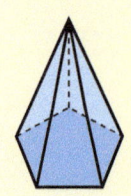   pentagonal pyramid

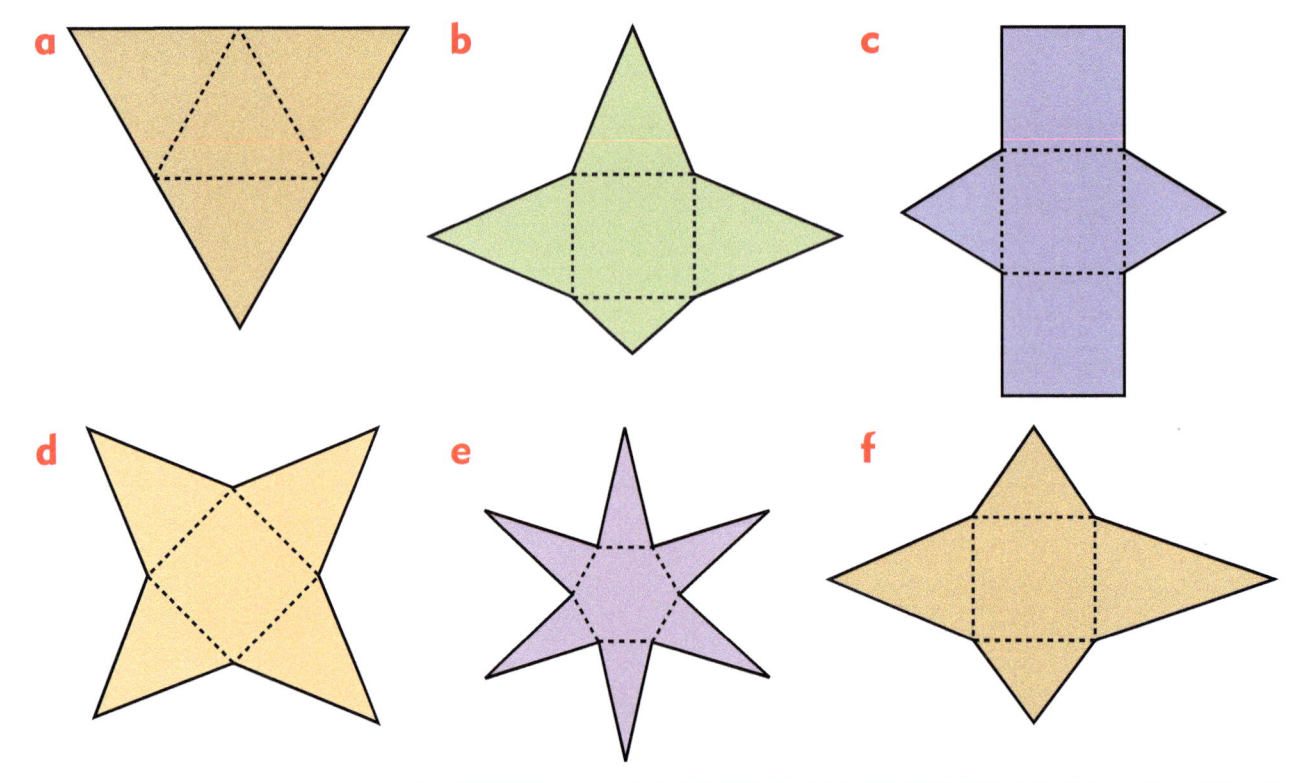

1 Say which of these shapes are nets of pyramids.

a   b   c

d   e   f

2 Draw each net, cut it out and try to make the shapes.

68

Negative numbers

Have you ever seen numbers with a minus sign in front of them?

This elevator has floors below the ground floor.

If you count down from the second floor, the floors go down in this order:

2, 1, 0, −1, −2

Numbers that are less than 0 are called **negative numbers**. They are written with a minus sign.

1 Copy each number line and complete the missing numbers.

a   −6 ___ −4 ___ ___ −1 0 1 2 ___ 4 ___ 6 7 8

b   −25 −24 ___ ___ −21 −20 −19

c   −150 −100 −50 ___ ___

d   −60 ___ ___ 20 40 ___

e   ___ ___ ___ 0 10 ___ 30

2 Copy and fill in <, > or = between each pair of numbers.

a   25 ___ −25
b   −5 ___ −6
c   0 ___ −5
d   −10 ___ −5
e   −2 ___ −1
f   −100 ___ −200

3 Did you have any difficulty comparing the numbers in question 2? Why? Write a tip for comparing negative numbers.

you can use Workbook page 39

Reading a thermometer

°C means 'degrees Celsius'.

Degrees are the units of measurement for heat.

0°C is the temperature at which water freezes (turns to ice). Negative numbers on a thermometer show us temperatures below freezing.

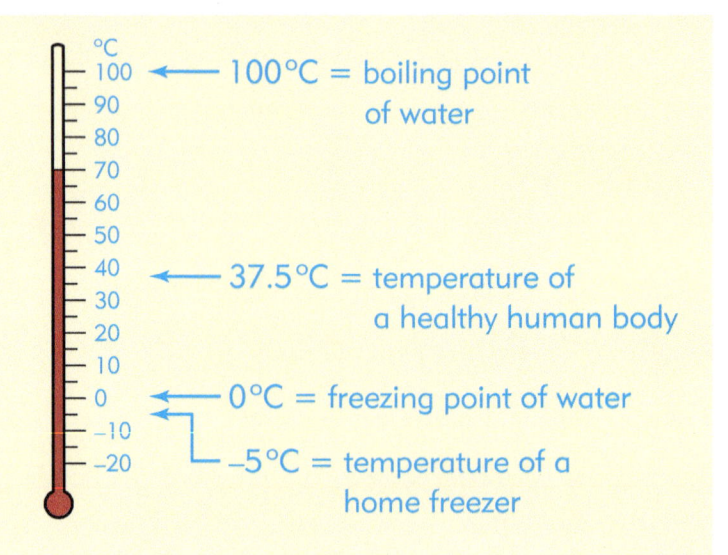

100°C = boiling point of water

37.5°C = temperature of a healthy human body

0°C = freezing point of water

−5°C = temperature of a home freezer

1 Write the temperature shown on each thermometer.

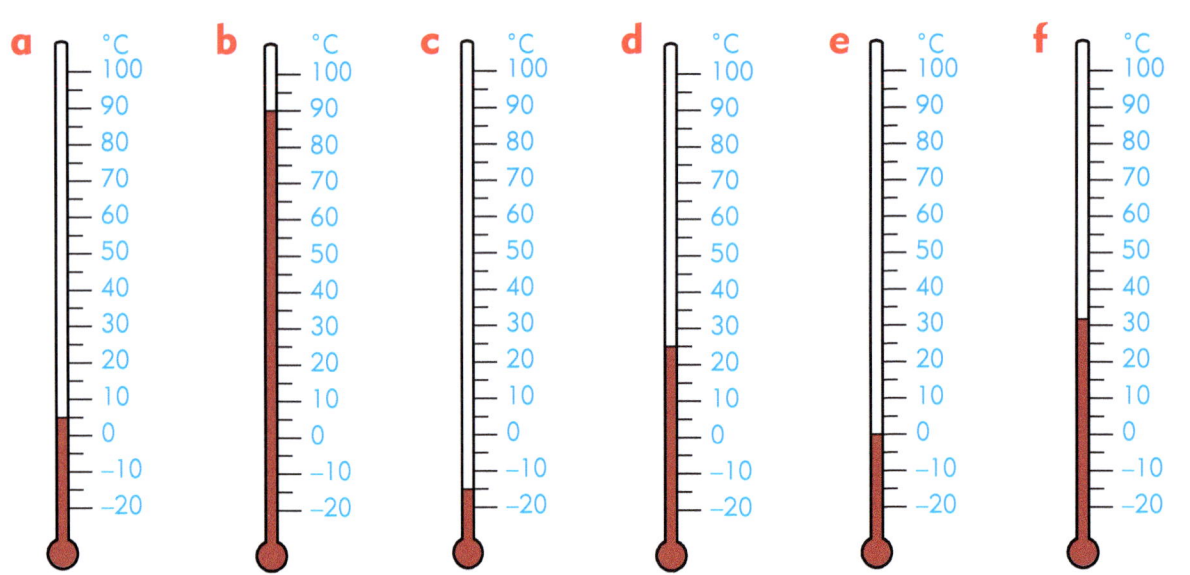

a   b   c   d   e   f

2 Write the temperatures from question 1 in order from coldest to hottest.

3 Find out the temperatures of these things.
  a The oven temperature for baking a cake.
  b The surface of Venus.
  c A car engine when it is running.

One day in winter

When Wendy went to sleep one winter's night, it was −5°C. During the night, the temperature fell 10°C to −15°C.

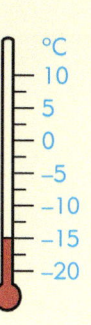

By noon the next day, the temperature had risen 20°C. It was then 5°C.

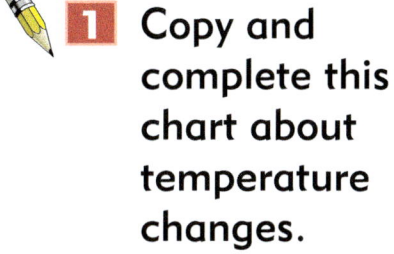

1 Copy and complete this chart about temperature changes. Estimate the time of day.

Starting temperature	Change	Temperature after change	Time of day
13°C	Rose 7°C		
8°C	Rose 5°C		
2°C	Fell 3°C		
−4°C	Fell 6°C		
−7°C	Fell 5°C		
−1°C	Rose 5°C		

2 Make your own temperature chart for a day. Take the temperature every hour.

3 Write down the warmest and coldest times of the day.

71

Fractions

Remember, a fraction is a part of an object or amount.
This shape is divided into four equal parts.
1 part of the 4 is shaded.
This is one-quarter. We write $\frac{1}{4}$.
3 parts of the 4 are not shaded.
This is three-quarters. We write $\frac{3}{4}$.
One-quarter plus three-quarters make the four-quarters or one whole shape.
We can write this as an addition sum: $\frac{1}{4} + \frac{3}{4} = 1$

You will need:
- squared paper
- scissors
- coloured pencils.

1 Cut out strips of squared paper and colour them to show these fractions.

a $\frac{1}{2}$ b $\frac{3}{4}$ c $\frac{2}{3}$ d $\frac{5}{8}$
e $\frac{4}{5}$ f $\frac{7}{8}$ g $\frac{3}{10}$ h $\frac{8}{10}$

2 Use your coloured strip to help you find the missing fractions in these sums.

a $\frac{1}{2} + \square = 1$ b $\frac{3}{4} + \square = 1$ c $\frac{2}{3} + \square = 1$

d $\frac{5}{8} + \square = 1$ e $\frac{4}{5} + \square = 1$ f $\frac{7}{8} + \square = 1$

g $\frac{7}{10} + \square = 1$ h $\frac{2}{10} + \square = 1$

3 Here is one way to colour one quarter of a square:

a Find as many other ways as you can to colour one quarter of a square.

b Choose a different fraction. Choose a different shape. Show your fraction in as many different ways as you can.

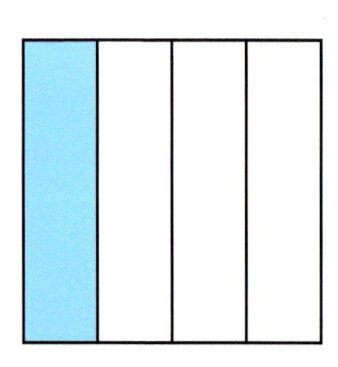

Fractions of shapes

This circle has been divided into two equal parts.
Each part is one half or $\frac{1}{2}$.
$\frac{1}{2}$ of the circle is coloured.
$\frac{1}{2}$ of the circle is not coloured.
$\frac{1}{2} + \frac{1}{2} = 1$ whole

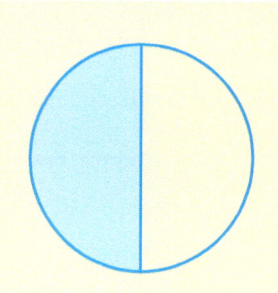

For each shape, write:

- **a** the number of parts it has been divided into
- **b** the fraction that is shaded
- **c** the fraction that is unshaded
- **d** the number sentence showing the sum of the shaded and unshaded parts, for example $\frac{1}{2} + \frac{1}{2} = 1$

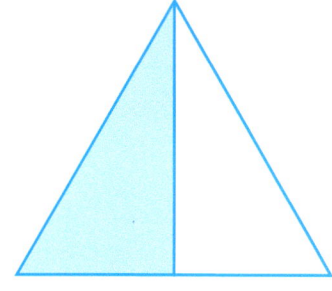

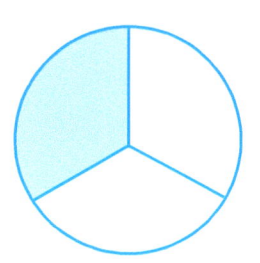

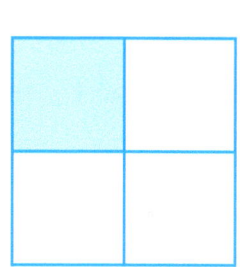

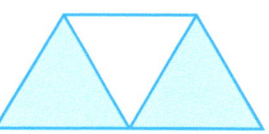

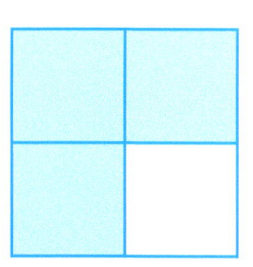

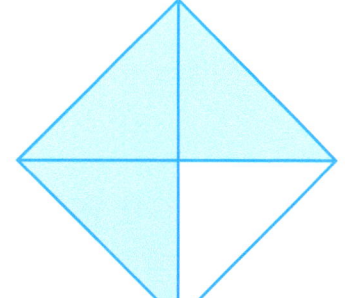

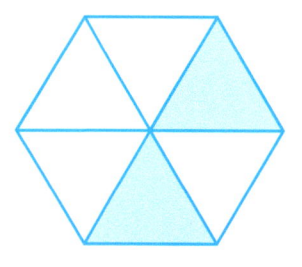

you can use Workbook page 40

Fractions of a number

This picture shows a big cube divided into a number of smaller cubes.

Use the picture or make your own cube to help you answer the questions.

1 Use your cube to calculate the following.

a How many cubes are there altogether?

b If you took one-third of the cubes away, how many would be left?

c Draw $\frac{1}{3}$ of the cubes.

d $\frac{2}{3}$ of the cubes are in one pile. The rest are in another pile. How many cubes are there in each pile?

2 These small cubes have been put together to make a cuboid.

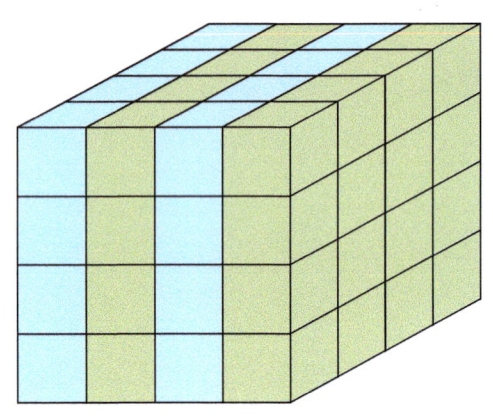

a How many cubes are there altogether?

b What fraction of the cubes are green?

c Draw $\frac{1}{8}$ of the blue cubes.

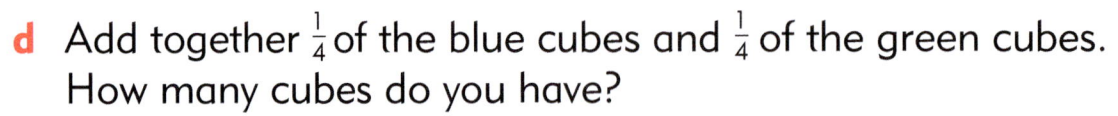

d Add together $\frac{1}{4}$ of the blue cubes and $\frac{1}{4}$ of the green cubes. How many cubes do you have?

e Put $\frac{1}{4}$ of the green cubes in a row. Put $\frac{3}{8}$ of the blue cubes in a row. Which row has more cubes?

3 Make or draw your own cuboid. Try a different size or use three colours for the bricks.

4 Make up some problems about your cuboid for a partner.

you can use Workbook page 41

Comparing fractions

Look at the shaded part of each shape.

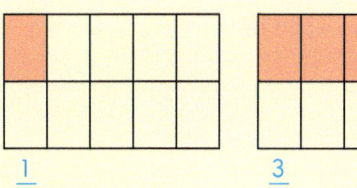

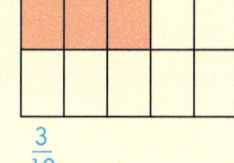

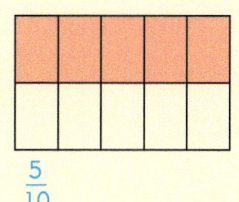

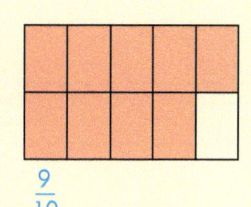

$\frac{1}{10}$ $\frac{3}{10}$ $\frac{5}{10}$ $\frac{9}{10}$

$\frac{1}{10}$ is less than $\frac{3}{10}$ You can write $\frac{1}{10} < \frac{3}{10}$
$\frac{9}{10}$ is more than $\frac{3}{10}$ You can write $\frac{9}{10} > \frac{3}{10}$

1 Copy the number sentences. Fill in < or > to compare the fractions.

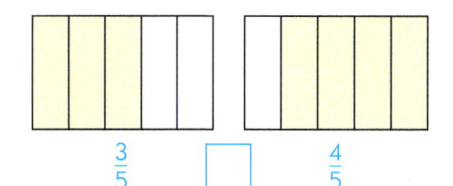

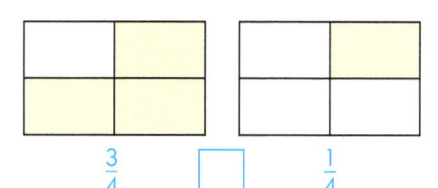

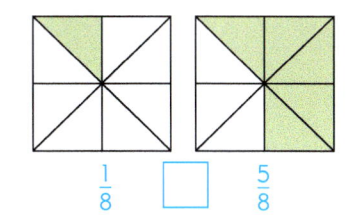

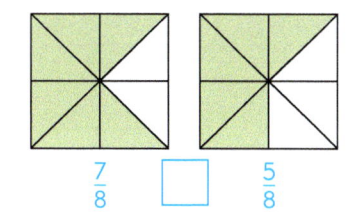

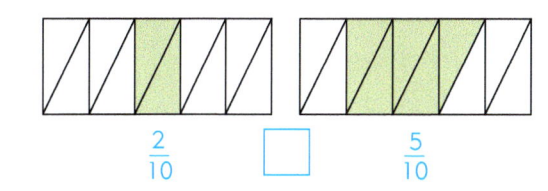

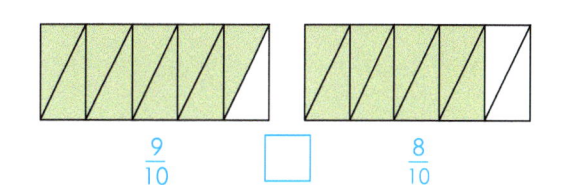

$\frac{3}{5}$ ☐ $\frac{4}{5}$ $\frac{3}{4}$ ☐ $\frac{1}{4}$

$\frac{1}{8}$ ☐ $\frac{5}{8}$ $\frac{7}{8}$ ☐ $\frac{5}{8}$

$\frac{2}{10}$ ☐ $\frac{5}{10}$ $\frac{9}{10}$ ☐ $\frac{8}{10}$

2 Write these fractions in order from biggest to smallest.

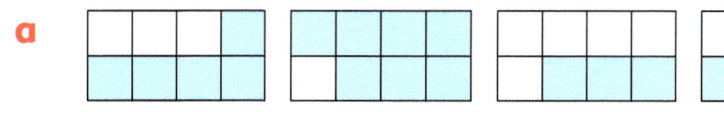

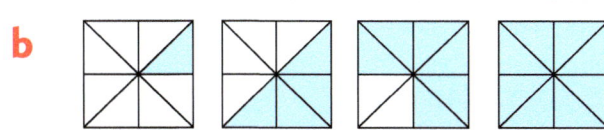

a

b

c

you can use Workbook page 42

Equivalent fractions

Use the fraction wall to help you compare fractions.
Fractions that are equal are called **equivalent fractions**.

$\frac{1}{2} > \frac{1}{3}$ $\frac{2}{4} = \frac{1}{2}$ $\frac{1}{4} < \frac{1}{3}$

1 Name four fractions that are equivalent to $\frac{1}{2}$

2 Which fraction from the fraction wall is equivalent to:

a $\frac{2}{10}$ b $\frac{4}{10}$ c $\frac{6}{10}$ d $\frac{8}{10}$?

3 What pattern do you notice in your answers to question 2?

4 Fill in <, > or =

a $\frac{1}{3} \square \frac{1}{4}$ b $\frac{3}{10} \square \frac{1}{5}$ c $\frac{3}{8} \square \frac{1}{2}$

d $\frac{3}{4} \square \frac{6}{8}$ e $\frac{4}{4} \square \frac{10}{10}$ f $\frac{3}{3} \square \frac{6}{6}$

5 Make up five of your own sentences showing equivalent fractions for the fraction wall.

you can use Workbook page 43

Compare and order equivalent fractions

1									
$\frac{1}{2}$					$\frac{1}{2}$				
$\frac{1}{3}$			$\frac{1}{3}$				$\frac{1}{3}$		
$\frac{1}{4}$		$\frac{1}{4}$			$\frac{1}{4}$			$\frac{1}{4}$	
$\frac{1}{5}$		$\frac{1}{5}$		$\frac{1}{5}$		$\frac{1}{5}$		$\frac{1}{5}$	
$\frac{1}{6}$	$\frac{1}{6}$		$\frac{1}{6}$		$\frac{1}{6}$		$\frac{1}{6}$		$\frac{1}{6}$
$\frac{1}{8}$	$\frac{1}{8}$	$\frac{1}{8}$	$\frac{1}{8}$	$\frac{1}{8}$	$\frac{1}{8}$	$\frac{1}{8}$	$\frac{1}{8}$		
$\frac{1}{10}$	$\frac{1}{10}$	$\frac{1}{10}$	$\frac{1}{10}$	$\frac{1}{10}$	$\frac{1}{10}$	$\frac{1}{10}$	$\frac{1}{10}$	$\frac{1}{10}$	$\frac{1}{10}$

Use the equivalent fraction wall to help you complete these activities.

1 Fill in <, = or > to make each number sentence true.

a $\frac{1}{2}$ ☐ $\frac{5}{10}$ b $\frac{6}{8}$ ☐ $\frac{3}{4}$ c $\frac{1}{5}$ ☐ $\frac{2}{10}$

d $\frac{5}{8}$ ☐ $\frac{1}{2}$ e $\frac{3}{4}$ ☐ $\frac{7}{10}$ f $\frac{1}{2}$ ☐ $\frac{5}{8}$

g $\frac{3}{4}$ ☐ $\frac{1}{2}$ h $\frac{7}{8}$ ☐ $\frac{3}{10}$ i $\frac{7}{8}$ ☐ $\frac{3}{4}$

j $\frac{1}{2}$ ☐ $\frac{1}{3}$ k $\frac{1}{2}$ ☐ $\frac{1}{5}$ l $\frac{1}{2}$ ☐ $\frac{3}{10}$

m $\frac{3}{5}$ ☐ $\frac{3}{10}$ n $\frac{4}{10}$ ☐ $\frac{1}{4}$ o $\frac{7}{10}$ ☐ $\frac{6}{8}$

2 Rewrite each set of fractions in order from smallest to greatest.

a $\frac{1}{2}$ $\frac{1}{4}$ $\frac{1}{3}$ $\frac{1}{10}$ $\frac{1}{8}$

b $\frac{2}{3}$ $\frac{2}{5}$ $\frac{2}{8}$ $\frac{2}{10}$ $\frac{2}{2}$

c $\frac{3}{5}$ $\frac{3}{8}$ $\frac{3}{10}$ $\frac{1}{2}$ $\frac{3}{4}$

d $\frac{5}{5}$ $\frac{5}{10}$ $\frac{5}{8}$ $\frac{1}{5}$

you can use Workbook page 44

Fractions and decimals

These strips are both divided into tenths.
We can write the fractions shown by each part as an ordinary fraction in tenths or as a decimal fraction using the decimal point.

$\frac{1}{10}$	$\frac{1}{10}$	$\frac{1}{10}$	$\frac{1}{10}$	$\frac{1}{10}$	$\frac{1}{10}$	$\frac{1}{10}$	$\frac{1}{10}$	$\frac{1}{10}$	$\frac{1}{10}$
0.1	0.1	0.1	0.1	0.1	0.1	0.1	0.1	0.1	0.1

$\frac{3}{10} = 0.3$

$\frac{5}{10} = 0.5$ but $\frac{5}{10}$ is also equivalent to $\frac{1}{2}$, so $0.5 = \frac{1}{2}$

These squares are divided into 100 equal parts. Each part is $\frac{1}{100}$th of the square. The shaded part of the square can be shown as an ordinary fraction and as a decimal.

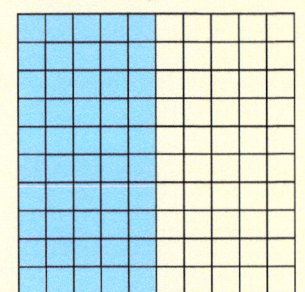

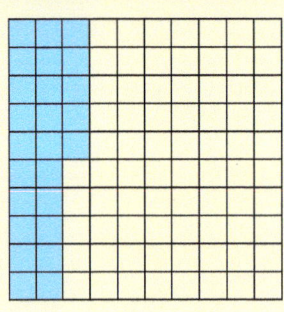

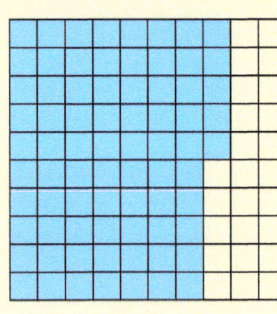

$\frac{50}{100} = 0.5$ $\frac{25}{100} = 0.25$ $\frac{75}{100} = 0.75$

$\frac{50}{100} = \frac{1}{2}$ $\frac{25}{100} = \frac{1}{4}$ $\frac{75}{100} = \frac{3}{4}$

$\frac{1}{2} = 0.5$ $\frac{1}{4} = 0.25$ $\frac{3}{4} = 0.75$

1 Write a decimal for each fraction.

a $\frac{1}{10}$ b $\frac{7}{10}$ c $\frac{75}{100}$

d $\frac{3}{10}$ e $\frac{5}{10}$ f $\frac{25}{100}$

2 Match the ordinary fractions with their decimal equivalents. Write your answers using equals signs like this: $\frac{1}{2} = 0.5$

a $\frac{8}{10}$ 0.5   b $\frac{1}{2}$ 0.25   c $\frac{1}{2}$ 0.8

    $\frac{3}{10}$ 0.8 $\frac{3}{4}$ 0.5 $\frac{9}{10}$ 0.5

    $\frac{5}{10}$ 0.3 $\frac{25}{100}$ 0.75 $\frac{80}{100}$ 0.9

you can use Workbook page 45

Mixed numbers

How many whole sandwiches can you make with these halves?

There are 5 halves. If we put them together we would have $2\frac{1}{2}$ sandwiches.

$2\frac{1}{2}$ is a mixed number. It is bigger than 2 and smaller than 3.

We can show $2\frac{1}{2}$ on a number line like this:

0    $\frac{1}{2}$ 1 $1\frac{1}{2}$ 2 $2\frac{1}{2}$ 3 $3\frac{1}{2}$

1 What fraction is coloured? Write the answers as mixed numbers.

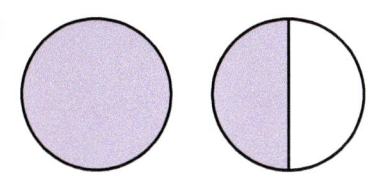

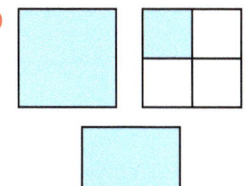

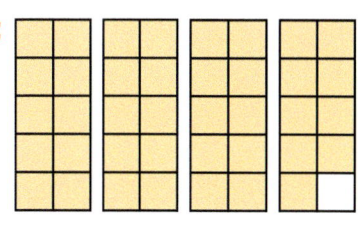

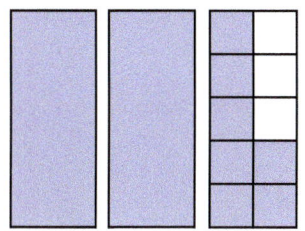

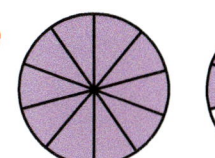

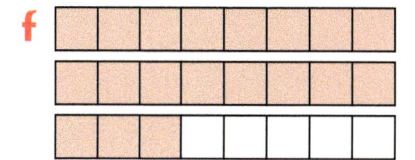

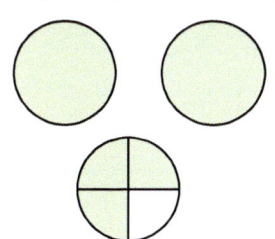

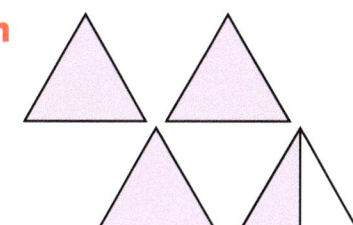

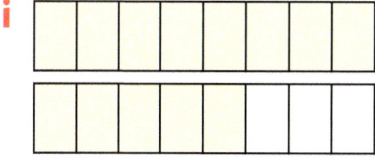

a   b   c

d   e   f

g   h   i

2 Draw number lines to show the position of each mixed number.

a $2\frac{1}{2}$ b $5\frac{3}{4}$ c $1\frac{2}{3}$

d $3\frac{1}{8}$ e $1\frac{1}{10}$ f $2\frac{3}{5}$

you can use Workbook page 46

Real-life problems

1 Solve these problems. Draw pictures to help you.

a The Stone family ordered 2 pizzas. Each pizza was cut into eighths. The family ate 15 pieces. What fraction of the pizzas did they eat?

b Steve picked some tomatoes from his garden. When he cut them into quarters he had 12 pieces. How many tomatoes did he pick?

c Four people ordered 5 chapattis with their curry. How much did each person get if they were shared exactly?

d Three cakes were cut up for the party. There were 15 guests and each had a fair share. What fraction was each cake cut into?

2 Write the mixed number.

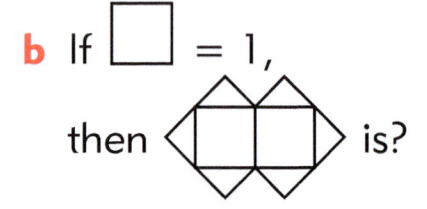

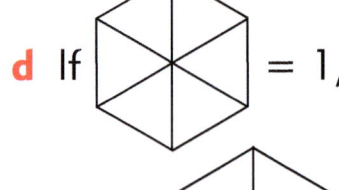

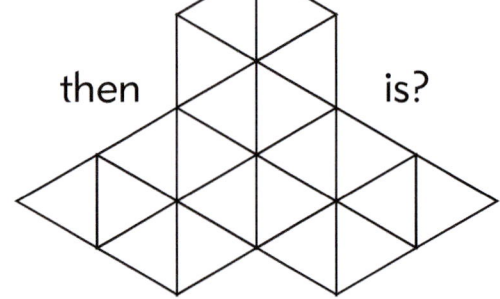

a If △ = 1, then [shape] is?

b If □ = 1, then [shape] is?

c If [square] = 1, then [shape] is?

d If [hexagon] = 1, then [shape] is?

3 Make up some similar puzzles for a friend to try.

80

Position on a grid

Look at this grid of shapes.

The triangle is in block D1.
The square is in block B2.

Some grids are labelled only with numbers.
To give the position of a square you have to give two numbers.

The numbers are called an ordered pair because they are always given in the same order.

- The number across is always given first.
- The number up is always given second.

On this grid, the rectangle is in block (2, 3).
The circle is in block (3, 2).

1 Write the colour and name of the shape in square:

a (2, 2) b (6, 1) c (1, 6)
d (4, 6) e (6, 4) f (5, 4)

2 Write an ordered pair to show the position of:

a the yellow circle
b the purple hexagon
c the pink parallelogram
d the green square
e the red rectangle
f the orange circle.

3 There is only one eight-sided shape on the grid.

a Find it and give its colour and correct name.
b What is the position of this shape?

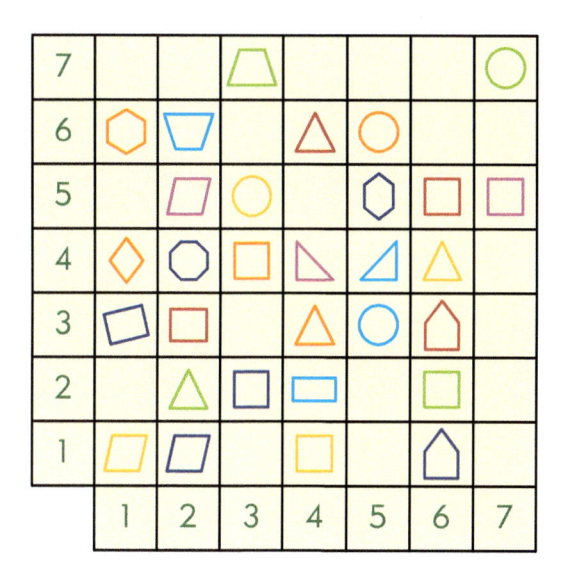

you can use Workbook pages 47–48

Compass directions

The Sun always comes up in the east. The Sun always goes down in the west.

East and west are points on a compass.

A compass has two other important directions. These are north and south.

The needle of a compass always points north.

The other directions are always in the same place compared to north.

We use the capital letters N, S, W and E to shorten the directions.

1 Here is a simple map of the main places in a town.

From the town centre:

a What is to the west?
b What is to the north?
c In which direction is the station?
d In which direction is the school?

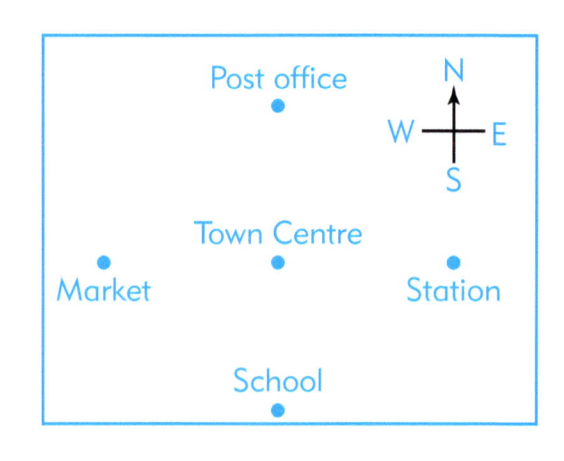

2 Work with a partner.

How many ways can you walk from the bus stop to the café?

Take turns to give each other directions from the bus stop to the café. Follow on the map as your partner tells you where to walk.

Write directions for the shortest route from the post office to the café.

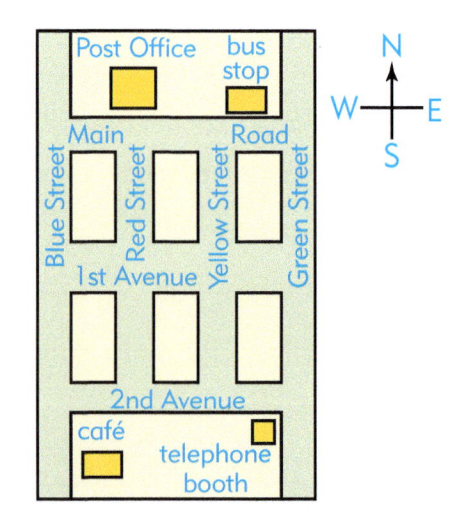

you can use Workbook page 49

82

Finding your way

The diagram shows how two counters have moved.
We can use directions to describe movement.
The red counter has moved 3 blocks south.
The blue counter has moved 2 blocks west and then 1 block north.
We write this as:

    2 west
    1 north

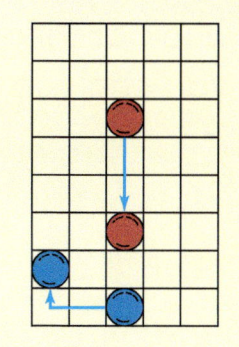

1 Write the directions followed by each counter.

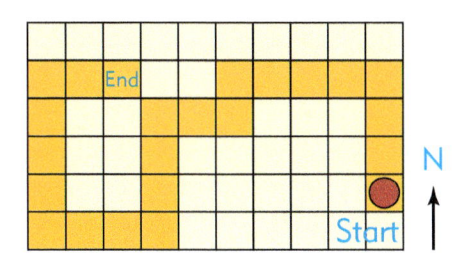

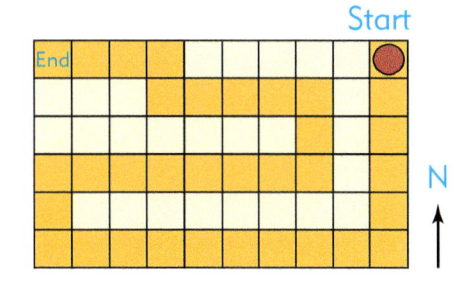

2 Only one path gets you to the tree.
  a Find the correct path.
  b Write directions from the gate to the tree.

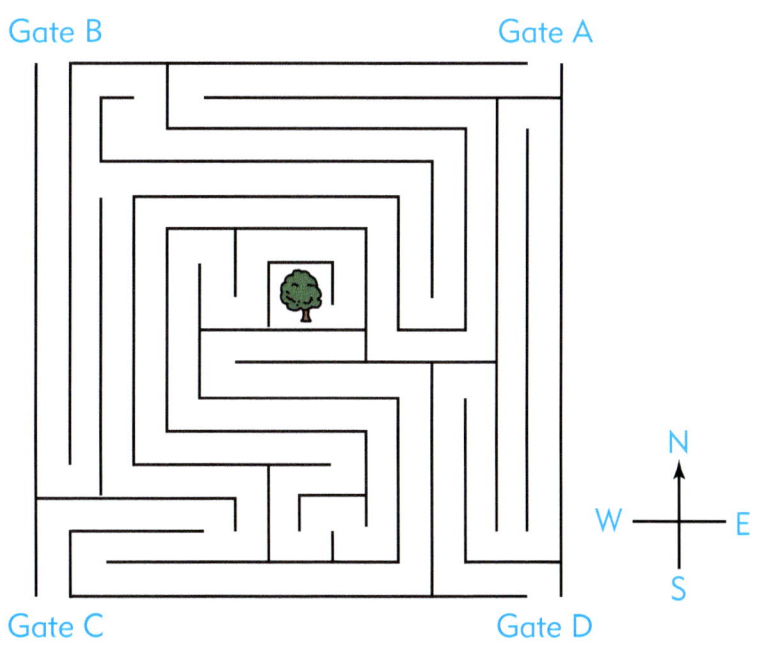

you can use Workbook pages 50–51

Mental strategies for adding

Look at these examples to see some of the strategies that you can use to add.

Counting on in steps
46 + 43
46 + 43 = 89

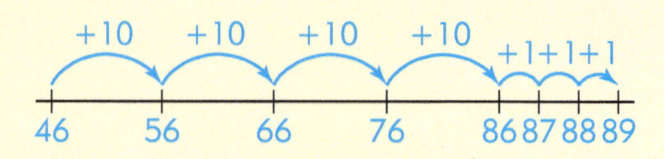

+10 +10 +10 +10 +1+1+1
46 56 66 76 86 87 88 89

Adding the nearest 10 or 100 and compensating
53 + 29 = 82   Because it is the same as 53 + 30 − 1
299 + 158 = 457   Because it is the same as 300 + 158 − 1

Partitioning the numbers using place value
35 + 42 = 30 + 5 + 40 + 2
This is the same as 30 + 40 + 5 + 2 = 70 + 7 = 77

Breaking up the numbers to bridge tens
35 + 48 = 30 + 5 + 40 + 8
This is the same as 30 + 40 + 5 + 8
= 70 + 5 + 8   If you don't know 5 + 8 you can
= 70 + 5 + 5 + 3   break the 8 into 5 + 3
= 80 + 3 = 83

1 Choose the best strategy to do these additions.

- a 47 + 19
- b 53 + 29
- c 35 + 39
- d 48 + 29
- e 76 + 39
- f 88 + 23
- g 74 + 58
- h 23 + 19
- i 91 + 25

2 Add these numbers using the method you find easiest.

- a 42 + 31
- b 35 + 54
- c 43 + 16
- d 16 + 72
- e 36 + 61
- f 57 + 43
- g 31 + 82
- h 19 + 83
- i 91 + 53

3 What number is 99 more than 43?
Tell your partner how you worked this out.

you can use Workbook page 52

84

Estimating

You can estimate what your answer is likely to be before you calculate.
The estimate helps you decide whether your answer is reasonable or not.
You estimate by rounding off numbers.

29 + 32   29 rounds to 30   32 rounds to 30
Estimate: 30 + 30 = 60
Actual answer: 30 + 32 − 1 = 62 − 1 = 61
The answer is close to the estimate, so it is reasonable.

1 Estimate by rounding off and then work out the answers to these sums.

a 87 + 94   b 43 + 67   c 58 + 63
d 85 + 99   e 54 + 28   f 32 + 76

2 These are the prices of four tins of paint.

WHITE $29
ROOF PAINT $51
RED RED RED $33.00
BLUE $45

Estimate how much money you would need to buy:
a two tins of white paint
b two tins of roof paint
c a tin of white and a tin of red
d two tins of blue paint
e a tin of roof paint and a tin of blue paint.

3 Work out the actual cost of buying the paint in Question 2.
Use your estimates to check that your answers are reasonable.

Counting on and back to subtract

Look at the methods two students used to work out 93 − 88

These two numbers are quite close, so I counted up from the smaller number and kept track on my fingers.

I decided to count back using a number line.

Start at 88
89 90 91 92 93
The difference is 5
so 93 − 88 = 5.

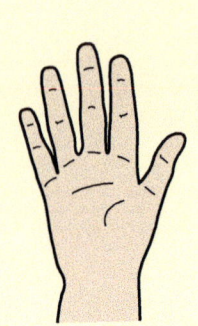

I started at 93
I jumped back 3 to get to 90
I jumped back 2 to get to 88
3 + 2 = 5 jumps,
so 93 − 88 = 5

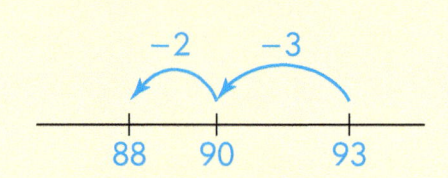

−2 −3
88 90 93

You can also count on or back in steps when you have to subtract a small number from a larger number. Look at this example to see how to subtract 403 − 9.

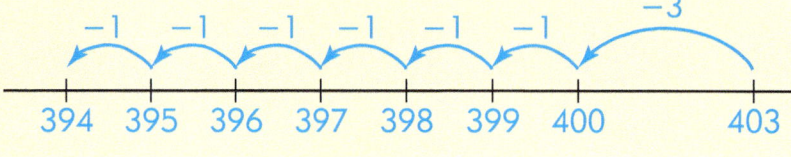

−1 −1 −1 −1 −1 −1   −3
394 395 396 397 398 399 400   403

1 Find the difference between each of these pairs of numbers.

a 34 − 29   b 48 − 39   c 87 − 79
d 84 − 78   e 45 − 39   f 42 − 37
g 65 − 59   h 75 − 69   i 53 − 47

2 Subtract. Use jottings to show any working you do.

a 203 − 9   b 304 − 8   c 107 − 9
d 406 − 8   e 501 − 9   f 402 − 5

More subtraction strategies

Read through these examples carefully.

Counting on or back in bigger steps
96 − 43
96 − 43 = 53

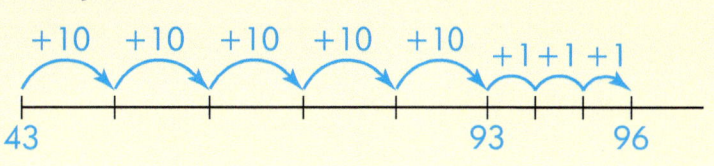

Making tens and hundreds to get easier numbers
73 − 28   Add 2 onto each number
75 − 30   75 − 30 = 75 − 10 − 10 − 10 = 45
So, 73 − 28 = 45

Using rounding and adjusting
147 − 29   29 is close to 30, and that is easy to subtract
147 − 30   147 − 10 − 10 − 10 = 117

But we have subtracted 1 too many, so we need to add it back 117 + 1 = 118. So, 147 − 29 = 118.

1 Subtract

a 56 − 19   b 73 − 29   c 41 − 19
d 89 − 87   e 63 − 20   f 80 − 48
g 72 − 39   h 33 − 18   i 77 − 49
j 61 − 25   k 75 − 17   l 74 − 38

2 Find the missing numbers.

a 53 − ☐ = 4   b 105 − ☐ = 7   c ☐ − 78 = 8
d 107 − ☐ = 8   e ☐ − 86 = 7   f 103 − ☐ = 9
g 96 − ☐ = 78   h 93 − ☐ = 59   i 91 − ☐ = 25

3 Show your partner how you worked out what the missing numbers were.

Working with bigger numbers

Here are three written methods you can use to calculate 238 + 175

Using place value

$$238 = 200 + 30 + 8$$
$$+185 = \underline{100 + 80 + 5}$$
$$300 + 110 + 13 = 423$$

Vertical method adding hundreds first

```
  238
 +185
  300
  110
   13
  423
```

Vertical method adding units first

```
  238
 +185
   13
  110
  300
  423
```

Here are two different ways of calculating 336 − 178

Vertical method showing totals at the side

```
  336
 −178
    2  →180
   20  →200
  100  →300
   36  →336
  158
```

Counting up in steps using a number line

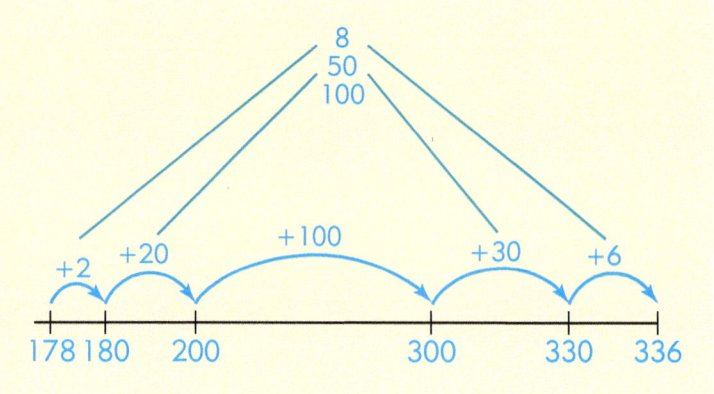

8
50
100

+2 +20 +100 +30 +6

178 180 200   300 330 336

1 Calculate. If you use a written method, show your working.

a 231 + 128   b 327 + 184   c 841 + 193   d 305 + 105
e 228 + 336   f 818 + 245   g 253 + 128   h 498 + 328

2 Subtract. If you use a written method, show your working.

a 692 − 545   b 321 − 267   c 816 − 797   d 433 − 312
e 924 − 799   f 211 − 158   g 572 − 451   h 734 − 686

you can use Workbook page 53

More adding and subtracting

1 Investigate subtraction and addition.

Work through this example:

Write down a three-digit number.	734
Reverse it.	437
Take away.	734 – 437 = 297
Reverse it.	792
Add.	297 + 792 = 1089

a Try a different three-digit number and do the same. What do you notice?

b Try one more. What happens?

c Try it with two-digit numbers. What do you notice about the answers?

2 I spilled coffee on this list of price changes. Write out a new list showing the price changes.

Item	Old price	New price	Price rise
Fridge	$325	$359	
Washer/drier	$427		$125
Cooker	$645		$254
Microwave	$197	$215	
Dishwasher		$385	$98

you can use Workbook page 54

Coded subtractions

In this code, each letter of the alphabet stands for a digit from 0 to 9:

A B C D E F G H I J K L M N O P Q R S T U V W X Y Z
0 1 2 3 4 5 6 7 8 9 0 1 2 3 4 5 6 7 8 9 0 1 2 3 4 5

1 Solve these problems using the code.
  a Subtract **ABC** from **DEF**. Write the number.
  b What number is the difference between **POD** and **PEA**?
  c Write a code for the answer to 471 – 289.
  d What number is the difference between **HIM** and **HER**?
  e Subtract **LMN** from **HIJ**. Write the number.
  f Write a code for the answer to 689 – 302.
  g What number is the difference between **FISH** and **CAT**?
  h Subtract **WXY** from **STU**. Write the number.
  i Write a code for the answer to 842 – 377.
  j What number is the difference between **CAR** and **BUS**?

2 Make up your own coded subtractions.

3 In these subtractions, some of the digits are missing. Write them out with all the digits.
  a 439 – 1__6 = 31__
  b 5__7 – 263 = __74
  c __25 – 532 = 1__3
  d 86__ – __45 = 623

Fruit and nut problems

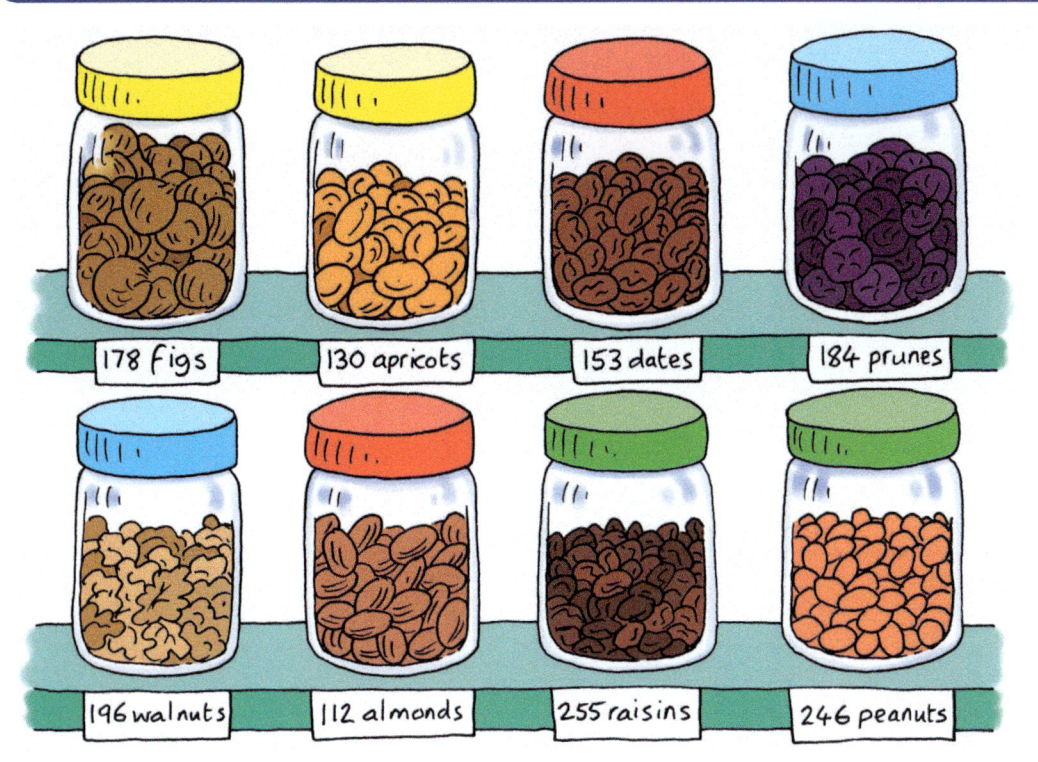

178 figs 130 apricots 153 dates 184 prunes
196 walnuts 112 almonds 255 raisins 246 peanuts

1 Use the numbers of fruits and nuts to calculate the following.

a How many figs and prunes are there altogether?
b How many more peanuts are there than almonds?
c How many apricots and figs are there altogether?
d What is the difference between the number of peanuts and raisins?
e I want to give my friend 200 dates. How many more will I need?

2 Use the numbers in the boxes. What do you need to add to each number to make 950?

246	353	537
883	902	364
518	672	826

you can use Workbook page 55

Perimeter

Perimeter is the distance around a shape.
This rectangle has a perimeter of 16 cm.
You can find this by adding the lengths of all the sides.

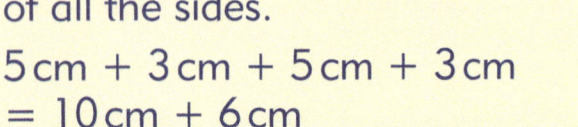

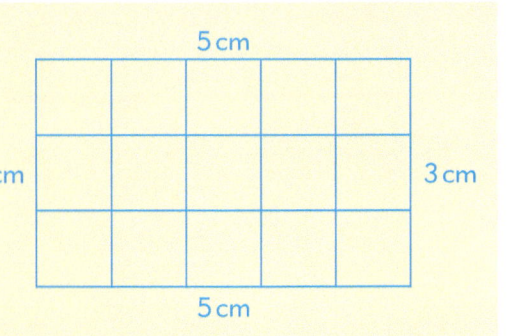

5 cm + 3 cm + 5 cm + 3 cm
= 10 cm + 6 cm
= 16 cm

The opposite sides of rectangles are the same length. So you can also find the perimeter by adding the length to the breadth and then doubling.

Length = 5 cm
Breadth = 3 cm
Length + breadth = 5 cm + 3 cm = 8 cm
Double 8 = 16 cm

1 Naadira cut these shapes out of 1 cm grid paper. Calculate the perimeter of each shape.

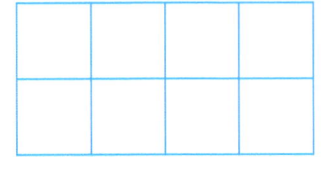

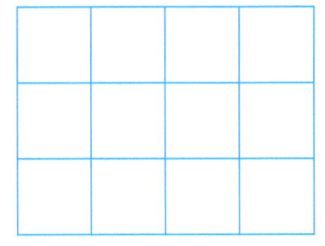

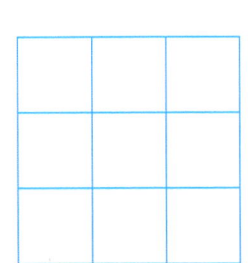

2 Draw these rectangles in your book. Write the perimeter of each one next to it.
  a 6 cm long and 3 cm wide
  b 9 cm long and 2 cm wide
  c 4 cm wide and 10 cm long
  d 1 cm wide and 3 cm long

you can use Workbook pages 56–57

Area

Area is the amount of space taken up or covered by a 2D shape.

This shape has been drawn on a grid of squares. If you count the square inside the shape you will see that it covers an area of 10 squares.

The squares on the grid all have sides that are 1 cm long.

A square with sides that are 1 cm long is called a square centimetre.

We can write this as 1 square centimetre or in a short way as 1 cm². The little 2 written above the unit means 'squared'.

The rectangle on the grid has an area of 10 cm².

1 Count the squares to find the area of each shape. Remember to write the units correctly.

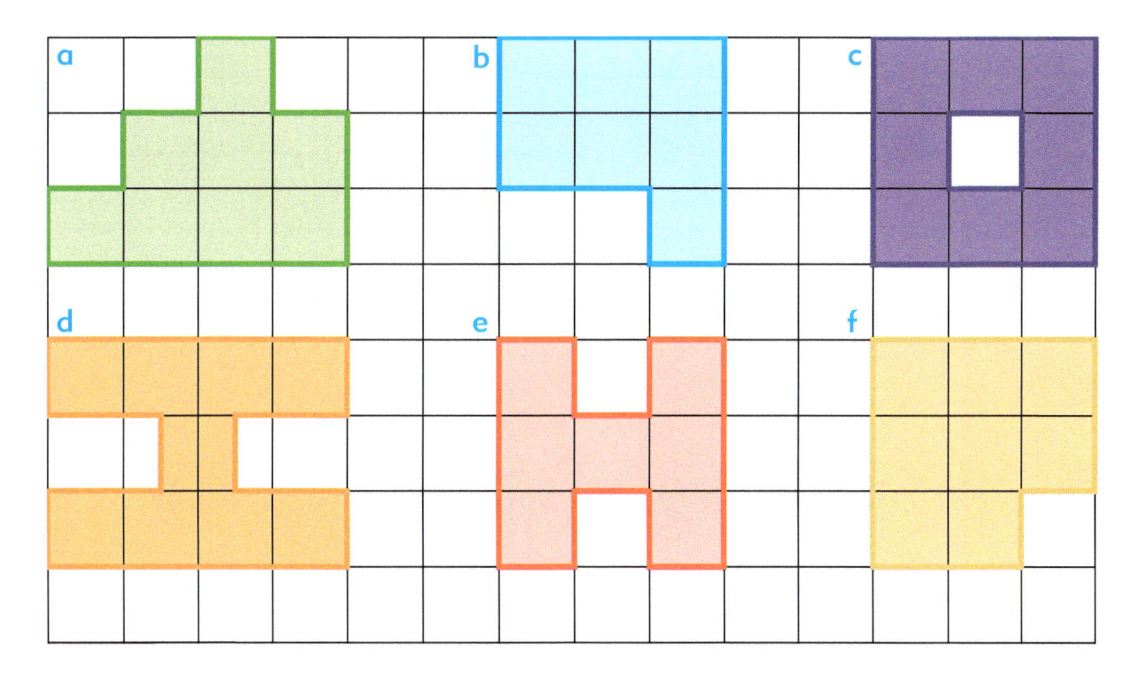

you can use Workbook pages 58–59

More area

1 Find the area of each shape in square centimetres.

2 Which shape covers:

a the greatest area

b the smallest area.

3 What is the total area of all the shapes?

Odd and even numbers

Even numbers can be arranged in groups of 2 with none left over. These are all even numbers:

10   14   22

Even numbers all have 0, 2, 4, 6 or 8 in the units place.

If odd numbers are arranged in groups of 2, there will always be one left over. These are all odd numbers:

11   15   23

Odd numbers all have 1, 3, 5, 7 or 9 in the units place.

1 **a** List the odd numbers.

19   36   88   103   199   432   555   9997   441   338

   b Write the next two odd numbers next to each one.

   c List the even numbers. Write the next two even numbers next to each one.

2 Use counters to work out whether your answer will be odd or even if you add the following:

  a odd + odd   **b** even + even   **c** odd + even   **d** even + odd

3 What happens when you subtract? Investigate whether your answer will be odd or even if you subtract the following:

  a odd − odd   **b** even − even   **c** odd − even   **d** even − odd

4 Tell the class what you discovered. Try to explain why this pattern works.

you can use Workbook page 60

Zig-zag number track

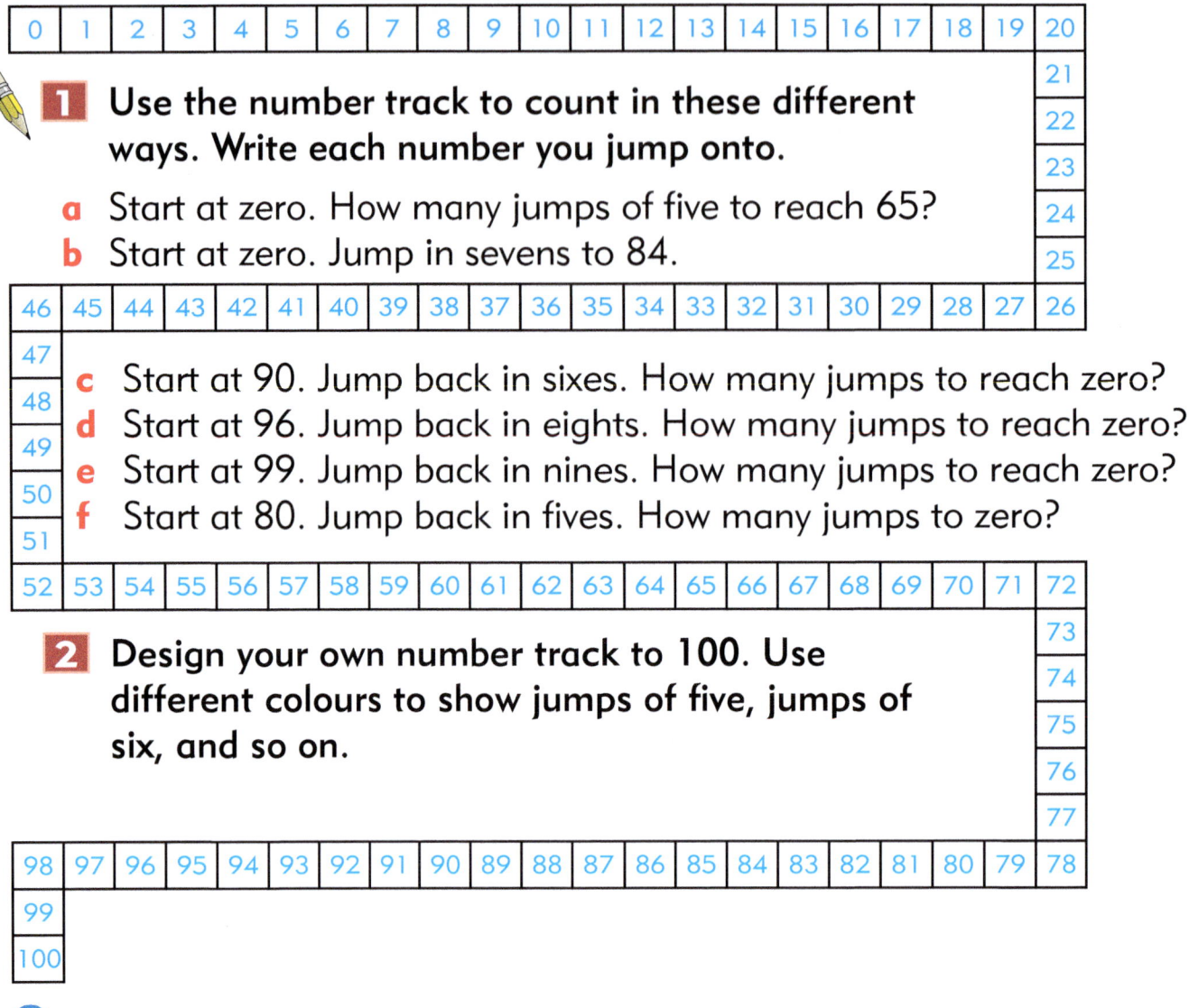

1 Use the number track to count in these different ways. Write each number you jump onto.

a Start at zero. How many jumps of five to reach 65?
b Start at zero. Jump in sevens to 84.
c Start at 90. Jump back in sixes. How many jumps to reach zero?
d Start at 96. Jump back in eights. How many jumps to reach zero?
e Start at 99. Jump back in nines. How many jumps to reach zero?
f Start at 80. Jump back in fives. How many jumps to zero?

2 Design your own number track to 100. Use different colours to show jumps of five, jumps of six, and so on.

3 Use the 0–100 track to work out the missing numbers in these patterns.

a __, __ 16, 20, 24, __, __
b __, __ 48, 42, 36, __ __
c __, __ 85, 80, 75, __ __
d __, __, 52, 56, 60, __, __

4 Can you complete these patterns?

a 125, 225, 325, __ __ __
b 450, 445, 440, __ __ __
c 321, 421, 521, __ __ __
d 1420, 1320, 1220, __ __ __

5 Describe how each pattern is made.

96

Number patterning

Cells increase by splitting.
You can use counters to represent cells.

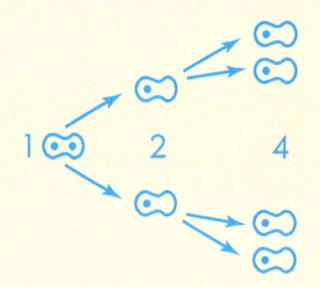

1 Start with one counter.

a Make a pattern to show how the cells continue to split.
b Draw the pattern and write the number of cells after each split. Carry on drawing as far as 6 splits.
c How many cells are there after 5 splits?
d How many cells would there be after 7 splits?
e How many cells would there be after 10 splits?
f What do you notice about the pattern?

2 Pauline used 1 cm mosaic tiles to make picture frames.

She wrote down how many tiles she used in each one. Her pattern was 10, 14, 18. She then measured the perimeter of each frame with a ruler. She found the measurements made a different pattern: 14 cm, 18 cm, 22 cm.

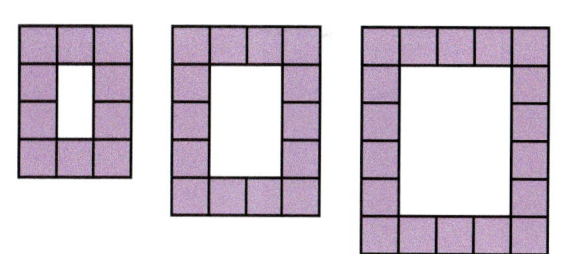

a Continue her pattern of picture frames until you have made six of them.
b Write down the numbers in your series of tiles and perimeters.
c How do the patterns grow?

you can use Workbook page 61

Multiples

A multiple is found by multiplying one number by another.
For example, when you multiply a number by 5, the answer is a multiple of 5.

1 × 5 = 5   2 × 5 = 10   3 × 5 = 15   4 × 5 = 20

5, 10, 15 and 20 are the first four multiples of 5.

1 Here are the first ten multiples of 5, 10 and 100.

Multiples of 5:   5, 10, 15, 20, 25, 30, 35, 40, 45, 50
Multiples of 10:   10, 20, 30, 40, 50, 60, 70, 80, 90, 100
Multiples of 100: 100, 200, 300, 400, 500, 600, 700, 800, 900, 1000

a Look at the multiples of 5. What digits are found in the units place?
b How can this help you recognise whether a number is a multiple of 5?
c Look at the multiples of 10. What digit is found in the units place?
d How can this help you decide whether a number is a multiple of 10?
e How will you know if a number is a multiple of 100?

2 Discuss these questions in your groups.

a If a number is a multiple of 10 it is also a multiple of 5. Why is this?
b Are all multiples of 5 also multiples of 10?
c Why are all multiples of 100 also multiples of 10?

3 What is the smallest number that is:

a a multiple of 5
b a multiple of 100
c a multiple of 5 and 10
d a multiple of 10 and 100?

you can use Workbook page 62

98

Angles

The size of an angle is a measure of the turning between the arms.

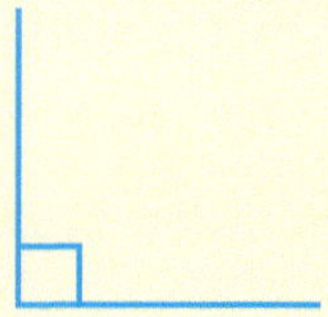

We measure angles in degrees.

One right angle = 90 degrees. We write this as 90°.

There are four right angles in a full turn, so there are 360° in a full turn.

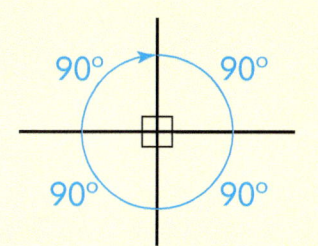

The length of the arms does not affect the size of the angle. Both these angles are the same size:

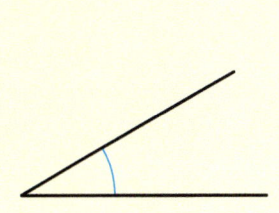

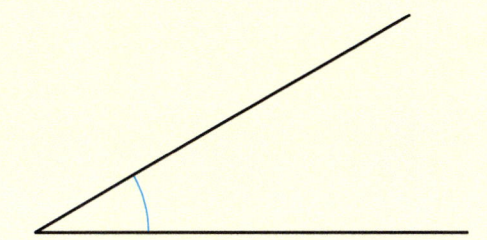

1 Look at the angles formed between the hands of these clocks.

a Which angles are right angles?

b Write the letters of the other angles in order from smallest to largest in size.

A B 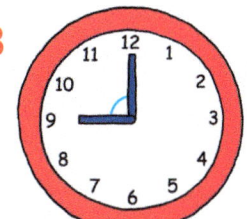 C

D 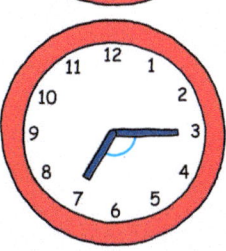 E  F

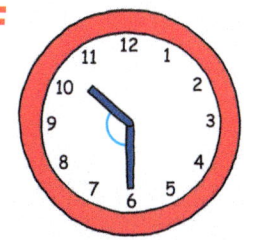

Compare and order angles

1 Make your own measuring tool for comparing the size of angles that look similar.
You will need:

- two strips of cardboard
- a paper-fastener.

Join the two strips at one end using the paper fastener like this:

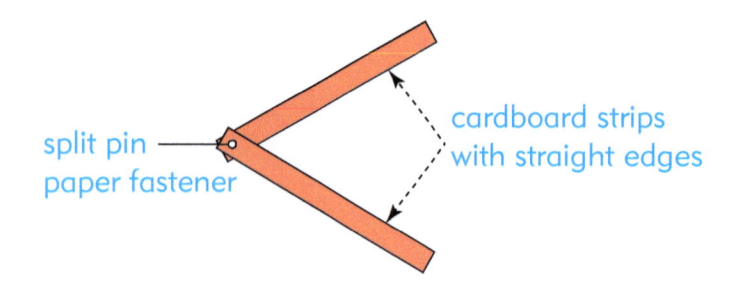

split pin
paper fastener

cardboard strips
with straight edges

The arms must be able to move, but they must not swing freely. Adjust the arms so that your angle is the same size as the one you are comparing.
Put the angle over another angle to see if it is smaller or larger than the one you measured.

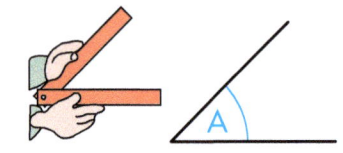

2 Arrange the angles in each set in order from greatest to smallest.

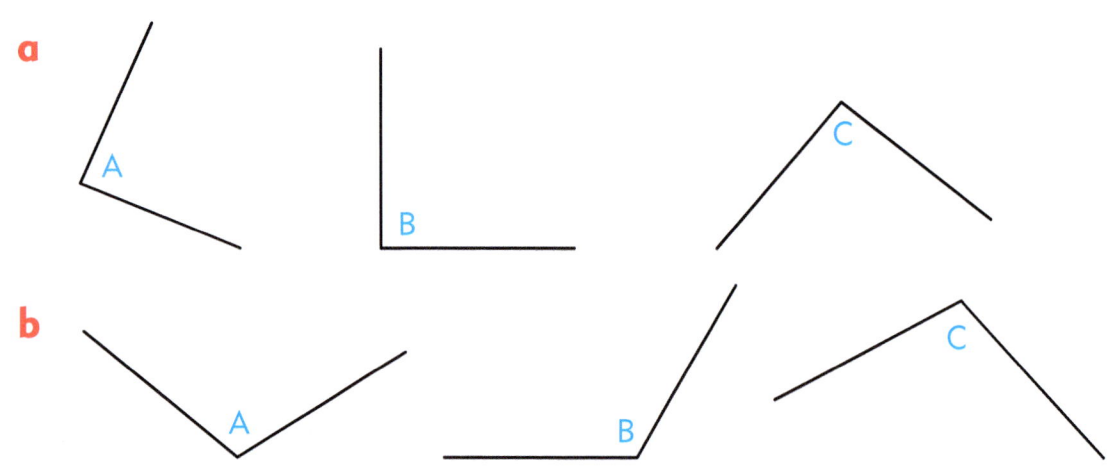

a
A   B   C

b
A   B   C

you can use Workbook pages 63–64

Revise multiplication facts

Nisha uses sets of matching cards like these to learn her multiplication facts.

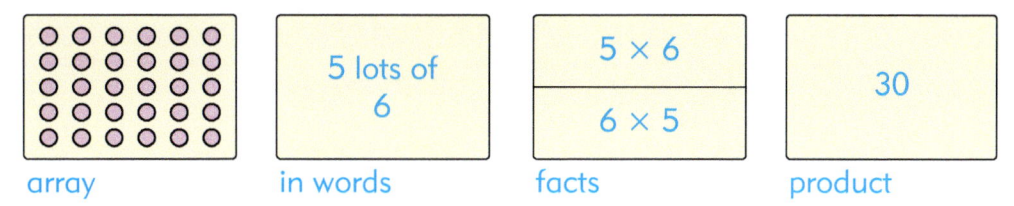

array	in words	facts	product
(6×5 array of dots)	5 lots of 6	5 × 6 / 6 × 5	30

These cards are all mixed up.

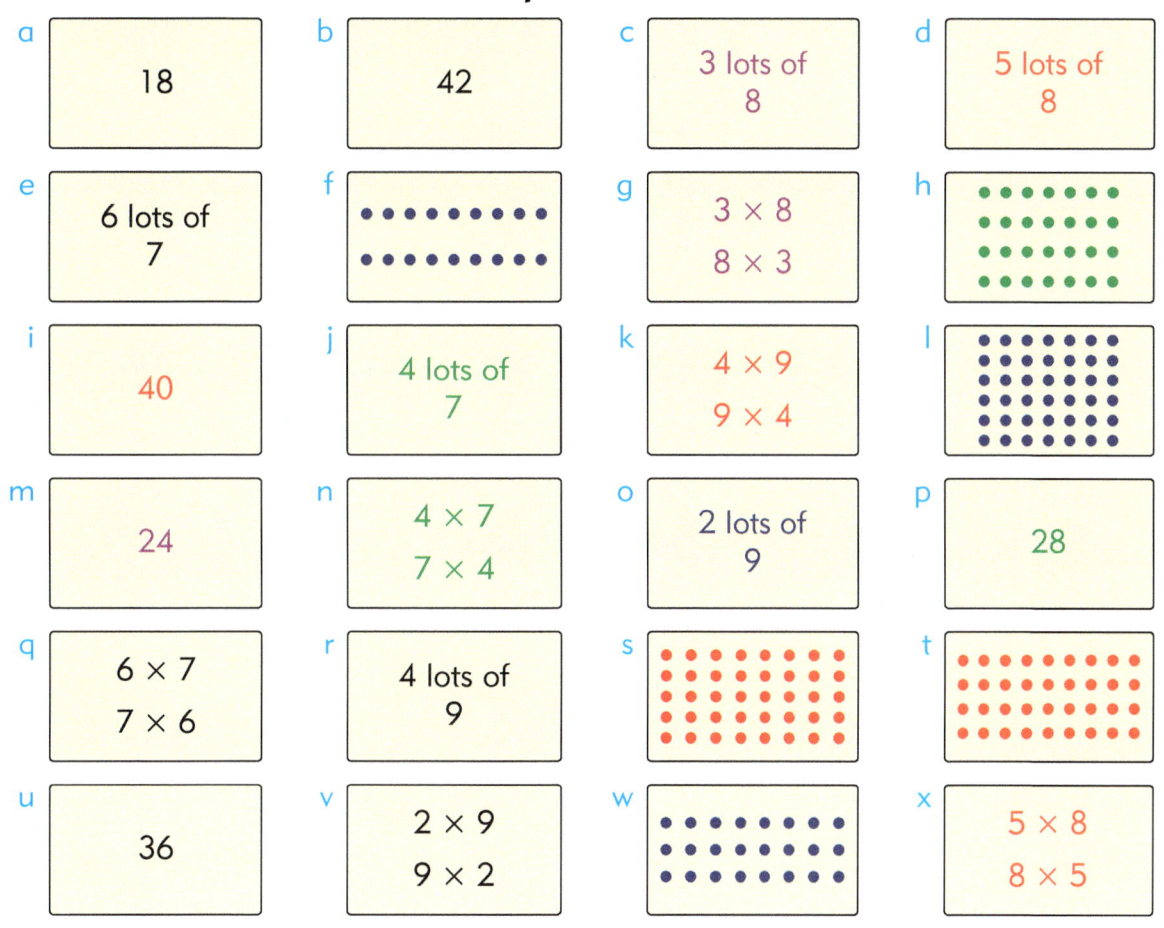

a. 18
b. 42
c. 3 lots of 8
d. 5 lots of 8
e. 6 lots of 7
f. (2×10 array of dots)
g. 3 × 8 / 8 × 3
h. (6×8 array of dots)
i. 40
j. 4 lots of 7
k. 4 × 9 / 9 × 4
l. (5×8 array of dots)
m. 24
n. 4 × 7 / 7 × 4
o. 2 lots of 9
p. 28
q. 6 × 7 / 7 × 6
r. 4 lots of 9
s. (5×8 array of dots)
t. (4×9 array of dots)
u. 36
v. 2 × 9 / 9 × 2
w. (3×8 array of dots)
x. 5 × 8 / 8 × 5

1 Work with a partner.
  a. Work out which cards go together.
  b. Write the letters of the cards that go together. You should have six sets each with four letters.

2 Make your own set of cards using different facts. Mix them up and give them to your partner to sort out.

you can use Workbook page 65

101

Multiplying tens

What is 50 × 3?
Rewrite this as 3 × 50 to make the multiplication easier.

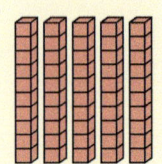

You know that 50 is five lots of 10

3 × 5 lots of 10 is 15 lots of 10
So, 3 × 50 is the same as 3 × 5 × 10
15 × 10 = 15 tens or 150
This can help you multiply any number by tens.

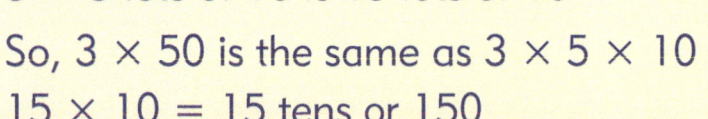

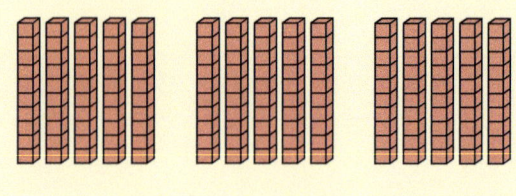

1 Write these amounts in numerals.
- a 7 tens
- b 9 tens
- c 11 tens
- d 21 tens
- e 28 tens
- f 30 tens
- g 50 tens
- h 72 tens
- i 99 tens

2 What is:
- a 6 × 3 tens
- b 3 × 9 tens
- c 4 × 5 tens
- d 5 × 6 tens
- e 3 × 3 tens
- f 9 × 10 tens

3 Calculate.
- a 6 × 20
- b 4 × 30
- c 5 × 60
- d 40 × 8
- e 50 × 7
- f 60 × 3
- g 80 × 9
- h 50 × 5
- i 90 × 4

4 80 planes leave an airport each day.
How many students can be carried in 9 buses?

5 A school bus can carry 60 students.
How many students can be carried in 9 buses?

Multiply bigger numbers by 10

You can use the multiplication facts and the patterns of multiples that you already know to find quick methods of multiplying any number by 10.

8 × 10 = 8 tens = 80
18 × 10 = 18 tens = 180
123 × 10 = 123 tens = 1230

We can show these numbers on place value tables:

Thousands	Hundreds	Tens	Ones
			8
		8	0

Thousands	Hundreds	Tens	Ones
		1	8
	1	8	0

Thousands	Hundreds	Tens	Ones
	1	2	3
1	2	3	0

When you multiply a whole number by 10, the digits move one place to the left on the place value table and you write a 0 as a place holder in the units place.

1 Write these amounts in numerals.

a 45 tens   b 67 tens   c 91 tens
d 146 tens   e 234 tens   f 420 tens

2 Try to do these multiplications mentally.

a 148 × 10   b 185 × 10   c 177 × 10
d 219 × 10   e 209 × 10   f 290 × 10
g 306 × 10   h 360 × 10   i 366 × 10

you can use Workbook page 66

Multiply by 100

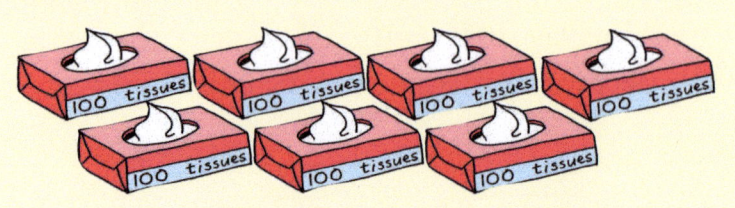

7 hundreds = 700, so
7 × 100 = 700

12 hundreds = 1200, so
12 × 100 = 1200

You already know that multiples of 100 have two zeros at the end. You can use this fact to quickly multiply any numbers by 100.

1 A pair of jeans costs $100. How much are the following?

   a 2 pairs of jeans
   b 3 pairs of jeans
   c 5 pairs of jeans
   d 7 pairs of jeans
   e 9 pairs of jeans
   f 11 pairs of jeans

2 Write these amounts in numerals.

   a 45 hundreds
   b 88 hundreds
   c 90 hundreds
   d 24 hundreds
   e 15 hundreds
   f 20 hundreds

3 Find the product of these numbers.

   a 3 × 100
   b 9 × 100
   c 12 × 100
   d 13 × 100
   e 25 × 100
   f 77 × 100

4 A tablet computer costs $99 on sale. Mrs Singh wants to buy 100 of these computers for her classroom. How much will they cost?

you can use Workbook page 67

Doubling

Do you remember how to double a number?

What is double 9?

9 + 9 = 18   or   9 × 2 = 18

Use the facts you know and place value to double bigger numbers.

What is double 38?
38 = 30 + 8
Double 30 = 60
Double 8 = 16
Double 38 = 76

What is double 130?
130 = 100 + 30
Double 100 = 200
Double 30 = 60
Double 130 = 260

What is double 1200?
1200 = 1000 + 200
Double 1000 = 2000
Double 200 = 400
Double 1200 = 2400

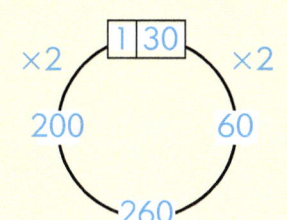

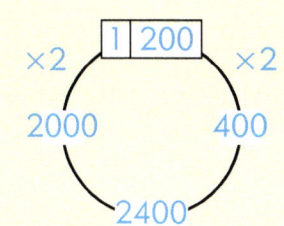

1 The weight of each item is given. What would two of the same item weigh?

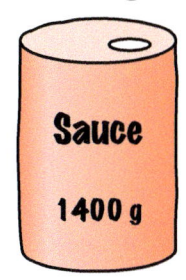

Chox 250 g   Sauce 1400 g   Tuna 85 g   Butter 370 g

2 Start with 9 and double five times. What do you get?

3 How many times do you have to double 10 before you get a result of more than 500?

you can use Workbook page 68

105

Halving

You already know that you can find half of a shape or amount by dividing it into 2 equal parts.

Halving is the inverse of doubling.

Double 9 = 18   Half of 18 = 9

You can use known facts or you can partition numbers to halve them easily.

What is half of 240?

Half of 24 = 24 ÷ 2 = 12
So half of 240 = 120

Halve 370
370 = 300 + 70
Half of 300 = 150   Think of this as half of 200 plus half of 100: 100 + 50
Half of 70 = 35   Think of this as half of 60 + half of 10: 30 + 5
Half of 370 = 185

1 Find half of each amount.

- a 28 km
- b 400 m
- c 1200 ml
- d $480
- e 4300 km
- f 56 mm
- g 150 km
- h 1500 m
- i 3700 ml

2 If you start on 1200, how many times must you halve to get to less than 100?

3 A travel agent is offering a half-price special on tours. Work out the new price of each tour.

- a Mumbai Magic – was $490, now ☐
- b New York by Night – was $1300, now ☐
- c Luxury London – was $330, now ☐
- d Sydney Surprise – was $3700, now ☐

you can use Workbook page 69

Multiplying a two-digit number by a one-digit number

Mrs Malone wants to make 15 of these 9-pin pinboards for her class.

How many nails will she need?

9 pins for each board times 15 boards.

We can work this out by multiplying.

Look at this example.

$15 \times 9 = (10 \times 9) + (5 \times 9)$
$= 90 + 45$
$= 90 + 10 + 35$
$= 135$

1 Find the product.

a 9×19
b 9×43
c 5×82
d 12×6
e 29×4
f 41×8
g 4×23
h 51×6
i 25×8

2 Calculate.

a 4×29
b 6×38
c 5×27
d 31×2
e 44×6
f 82×3
g 91×6
h 41×5
i 83×5

3 Millie works out 63×4 like this:

$63 + 63 = 126$
$126 + 126 = 240 + 12 = 252.$

a What did Millie do?
b When will this method be useful? Why?

you can use Workbook pages 70–71

Multiplication problems

1 A packet of apples contains 25 apples. How many apples would there be in 9 packets?

2 Sam's school is 29 km from his home.
 a What is the distance to school and back?
 b How far does Sam travel if he goes to school and back 5 times a week?
 c How far does Sam travel in two weeks?

3 43 children each have 6 school books. How many books is this altogether?

4 Seven 42-seater buses were needed for a school trip. How many seats is this altogether?

5 Nick ordered 8 boxes of biscuits for the tuck shop. Each box contains 48 packets of biscuits. How many packets of biscuits is this altogether?

6 Salma orders 57 sets of pens. Each set contains one red, one blue and one black pen. How many pens are in 57 sets?

7 75 students each paid $3 for bus fare and $5 for lunch and entry tickets on a school outing.
 a How much money did the teacher collect altogether?
 b What is double this amount?

8 At a school assembly there are 7 rows of 45 chairs and 10 rows of 55 chairs in the hall. How many chairs are there altogether?

Venn diagrams

Do you remember how to use a Venn diagram to sort information or objects?

This Venn diagram was used to sort shapes.

There are two groups: blue shapes and shapes with right angles. These are sorted into two circles.

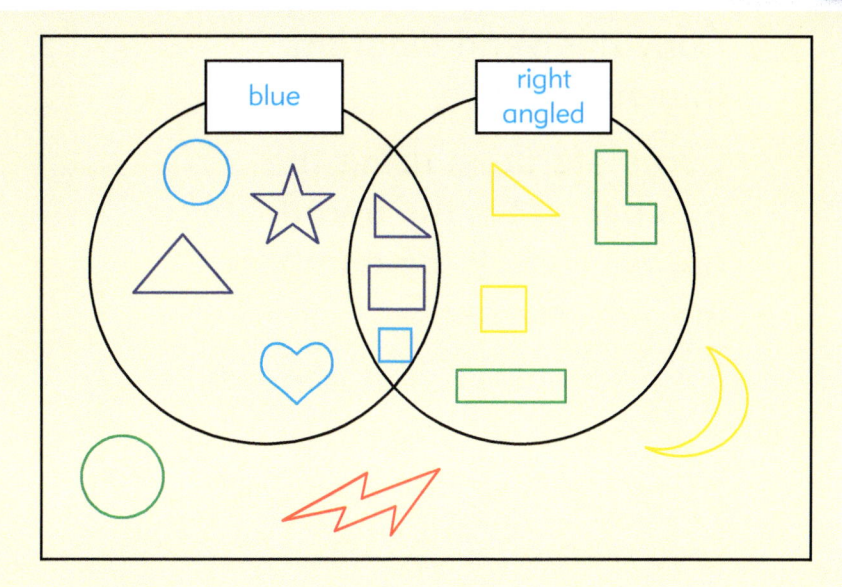

Some shapes fit into both groups. They go into the overlapping section of the circles.

Some shapes are not blue and they don't have any right angles. They go outside the circles.

1 Sam drew this Venn diagram to show which of her friends like Maths and which like English.

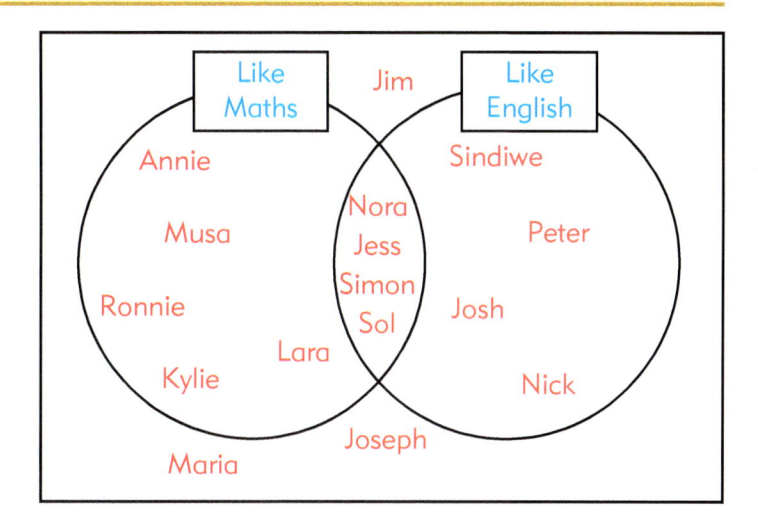

  a How many of her friends like Maths but not English?
  b How many like English, but not Maths?
  c How many like English and Maths?
  d Which friends do not like English or Maths?
  e How do you know which friends do not like Maths or English?

2 Do a quick survey of at least six of your own friends to find out who likes Maths and who likes English. Draw a Venn diagram to show your results.

you can use Workbook page 72

More Venn diagrams

1 Copy this Venn diagram four times.

Complete them using the numbers from the board above. Use all the numbers in each diagram.

0 1 2 3 4 5 6 7 8 9
10 11 12 13 14 15 16
17 18 19 20

A Even numbers   Numbers > 15
B Even numbers   Multiples of 3
C Multiples of 4   Numbers < 12
D Odd numbers   Numbers > 11

2 Make up one more Venn diagram of your own using these numbers. Draw it in your book.

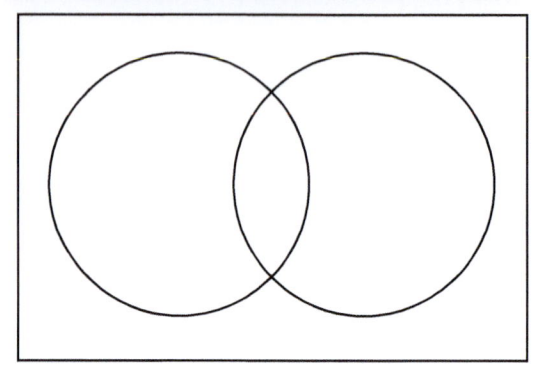

3 These two Venn diagrams are incorrect. Find the mistakes in each one and redraw them correctly.

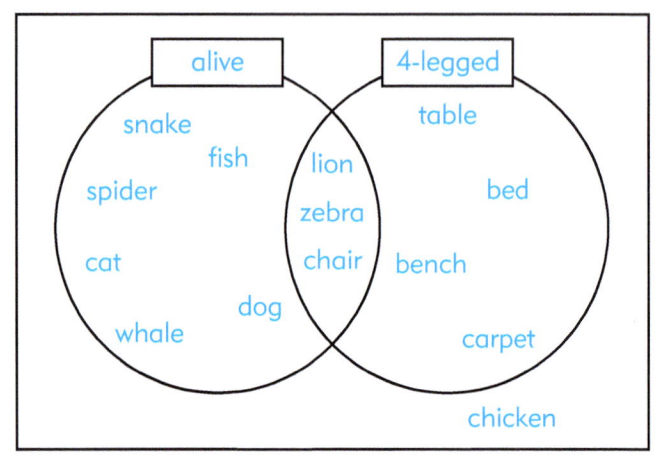

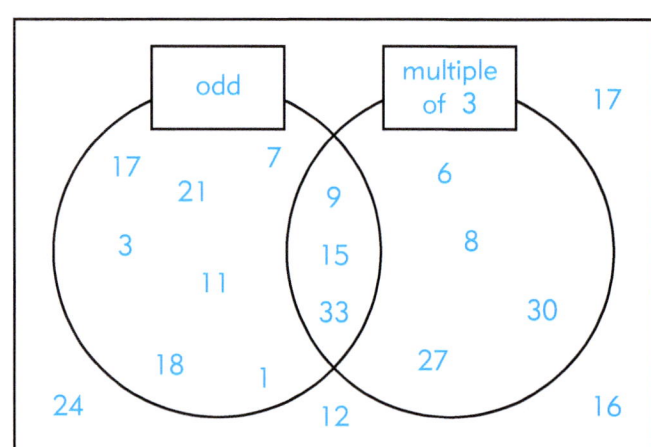

110

Carroll diagrams

Sofia made this chart about the children in her class. It is called a Carroll diagram.

	Boy	Not boy
Has brown eyes	John Mahmoud David Kyle	Jenny Ella Maria Yasmina
Does not have brown eyes	Adam Ismael Ed Tony	Sofia Chloë

You will need:

- cardboard
- tape or string
- sticky tac
- pens
- small cards or soft strips of paper.

1 Take a big piece of cardboard. Stick two pieces of coloured tape or string on your board to make a big Carroll diagram. Leave space at the top and on the left for labels. Make strips of paper with all the names of the people in your class.

a Make four cards:

| Boy | Not boy | Has brown eyes | Does not have brown eyes |

Stick the cards in the correct positions on the diagram.

Then arrange the name strips on the board.

b Think of different ways of sorting the children in your class. For example:

- boys and girls who do and don't wear glasses
- boys and girls who do or don't like sport.

you can use Workbook page 73

Sorting data into three groups

Look at this Venn diagram carefully.

There are three groups: blue shapes, small shapes and triangles.

Some shapes are small and blue. There are drawn where the blue and the small circles overlap.

Some shapes are small and triangles. These are drawn where the small and the triangles circles overlap.

Some shapes are blue and triangles. These are drawn where the blue and the triangles circles overlap.

Only one shape is small and blue and a triangle. It is drawn in the part of the diagram where all three circles overlap.

Two shapes are not blue, not small and not triangles. These are drawn outside the overlapping circles.

112

Sorting data into three groups *continued*

1 Answer the questions about this Venn diagram.

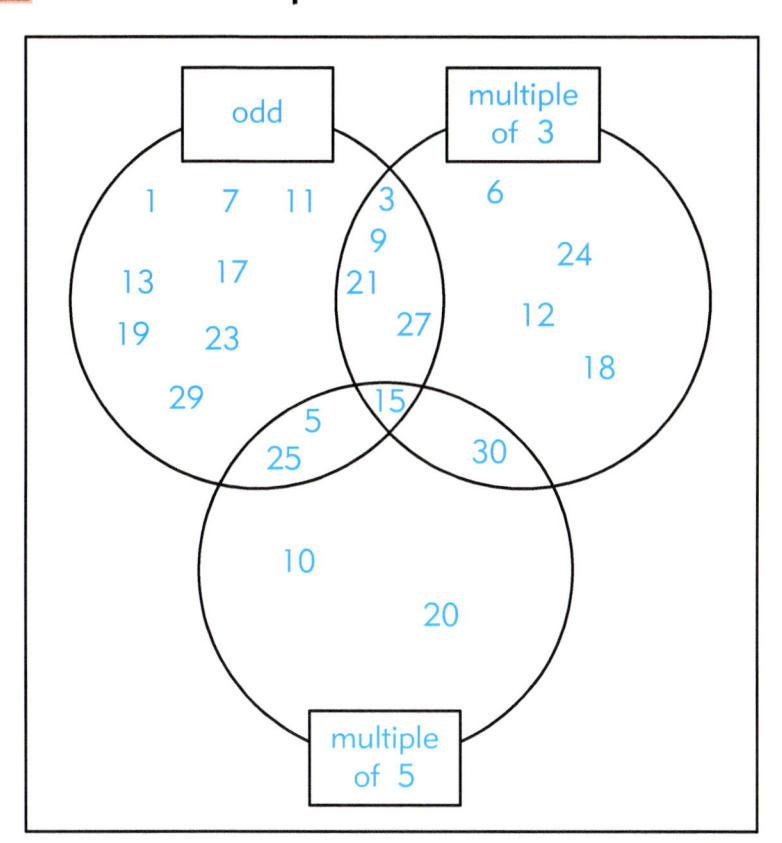

[Venn diagram with three overlapping circles labeled "odd", "multiple of 3", and "multiple of 5":
- odd only: 1, 7, 11, 13, 17, 19, 23, 29
- odd ∩ multiple of 3: 3, 9, 21, 27
- multiple of 3 only: 6, 24, 12, 18
- odd ∩ multiple of 5: 5, 25
- odd ∩ multiple of 3 ∩ multiple of 5: 15
- multiple of 3 ∩ multiple of 5: 30
- multiple of 5 only: 10, 20]

a Into which three groups were these numbers sorted?
b Which numbers are odd and multiples of 3, but not multiples of 5?
c Which numbers are even multiples of 3?
d Which odd numbers are multiples of 5, but not multiples of 3?
e What can you say about the number 15?

2 Work with a partner. Discuss where you would place the following numbers on the Venn diagram above. Give a reason for each decision.

a 33   b 35   c 40   d 44

e Try to find at least one other number that would fit into the same section as the 15.

113

Using Carroll diagrams to sort data

1 Copy this Carroll diagram. Write the letters of the shapes in the correct places in the diagram.

	Red	Not red
Quadrilateral		
Not quadrilateral		

2 Which shapes are:
   a red quadrilaterals?
   b quadrilaterals that are not red?
   c neither red, nor quadrilaterals?

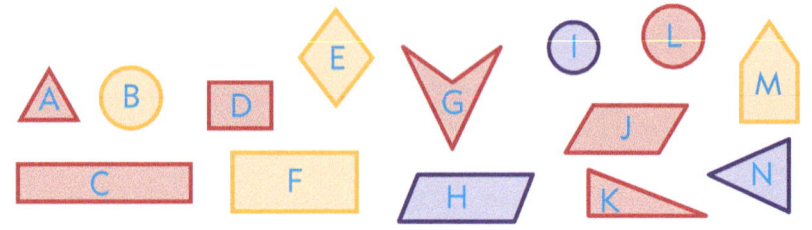

3 Copy this Carroll diagram. Complete it for numbers from 1 to 24.

	Even	Not even
Multiple of 4		
Not multiple of 4		

4 Do a quick survey among ten of your friends to find out who likes bananas and who likes raisins. Record your results on a Carroll diagram like this one.

	Like bananas	Don't like bananas
Like raisins		
Don't like raisins		

5 Complete these statements based on your completed diagram.
   a Most of my friends like …
   b More people like … than …
   c … don't like bananas or raisins.
   d I fit into the group that …

A database: big cats

A **database** is a chart or table of information.

It shows information set out in 'fields'. In this database about big cats, the fields are in the column at the side. They are 'length', 'continent', 'main prey', 'colour', and so on.

	Lion	Tiger	Leopard	Ocelot
Length	2.75 m	3 m	2.25 m	1.25 m
Continent	Africa	Asia	Asia, Africa	South America
Main prey	wildebeest	deer	baboons, deer	small mammals
Colour	pale brown (tawny)	orange and black	yellow and black	grey, fawn black
Markings	none	stripes	spots	spots and stripes
Habitat	plains	jungle	plains	forest and jungle

1 Use this Carroll diagram to sort the animals.

	Has spots	Does not have spots
Found in Africa		
Not found in Africa		

Using a database

1 Answer the following questions by using the 'big cats' database on the previous page.

a What do lions eat?

b How long is an ocelot?

c On which continent do tigers live?

d What sort of markings do leopards have?

e Which big cats do baboons avoid?

f Which cats live on the plains?

g Which is the longest big cat?

h Which big cat lives in South America?

2 Find out more about other big cats – for example, the panther, the lynx, the cheetah, etc.

Make your own database using the files and fields on the previous page. Add any more fields you are interested in – for example, life span, number of cubs in a litter, whether they are endangered or not.

3 Do you have a hobby, collection or special interest? Design your own database – for example, if you collect coins from different countries, you could write down the country, the currency, the value and what is shown on them.

More sorting

1 W = {days of the week}

F = {days or months starting with F}

M = {months of the year}

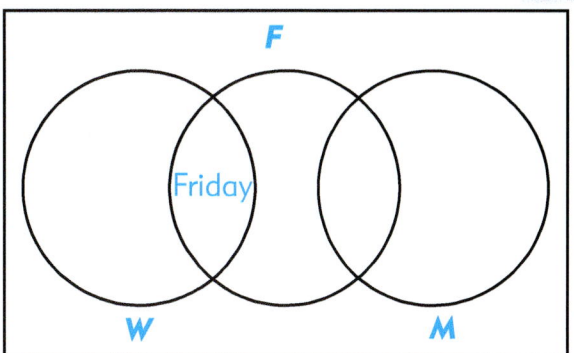

[Venn diagram with three overlapping circles labelled W, F, M; "Friday" is in the overlap of W and F]

Copy and complete the Venn diagram by filling in all the elements of sets W, F and M in the correct places.

2 Look at this Carroll diagram showing the number of T-shirts sold at a shop over a weekend. Answer the questions that follow.

a What are the four categories shown on this Carroll diagram?

b In which category would a green T-shirt with a picture of a dinosaur belong?

c Which was the most popular category?

d Which were more popular: T-shirts with words or T-shirts without words?

e Which were more popular: white T-shirts or T-shirts in other colours?

Types of T-shirt sold at a shop

	Has words	Does not have words
White	||||| |||||	|||
Not white	||||| ||||| ||||| |||	||||| ||||| ||

Dividing by sharing

Divide the fish into groups of 6.
We can make five groups of 6.
There are two fish left over.
32 ÷ 6 = 5 remainder 2
We write this as 5 r. 2

1 Can you split the group of fish into groups of 5?
Write out the division calculation, with the remainder.

2 Try to split the 32 fish into groups of:
a 2   b 3   c 4
d 7   e 9   f 10

Write each division with a remainder if necessary.

3 Try this with the number of students in your class.
a Split the total class into groups of 2, 3, 4, 5, 6, 7, 8, 9, 10.
b Write each division calculation with a remainder if necessary.

Division by repeated subtraction

What is 39 ÷ 5?

You can work this out by subtracting fives.

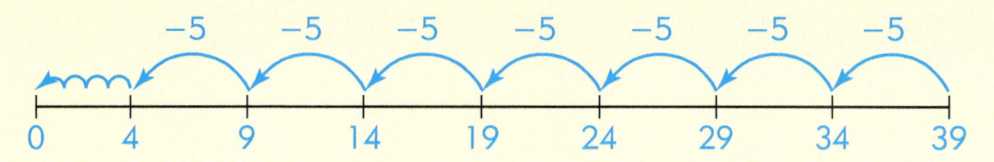

$39 \div 5 = 7$ remainder 4

You can do this in a shorter way like this:

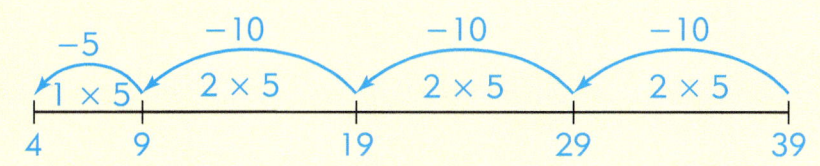

You can also do this by counting forwards using your multiplication facts.

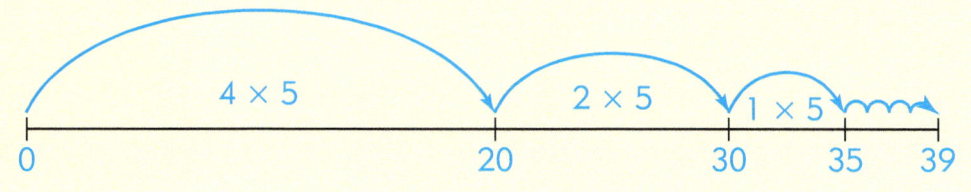

This is a shorter way of recording what you do.

39 ÷ 5
39 − 30 = 9   (5 × 6)
9 − 5 = 4   (5 × 1)
7 r 4

Divide the beads in each jar into 5 groups.

Write how many beads are left over.

a 19
b 42
c 51
d 67
e 79
f 83
g 99

you can use Workbook page 74

More dividing

You can use the multiplication facts you already know to help you divide.

How many groups of 10 can you make with 65 cubes?

10 × 6 = 60

10 × 7 = 70 This is too much, so the answer must be 6 groups with 5 left over.

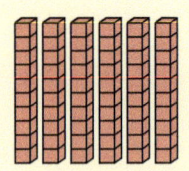

6 groups of 10   5 left over

1 Divide. Use your tables to help you.

a 20 ÷ 3   b 42 ÷ 10   c 63 ÷ 9
d 40 ÷ 6   e 43 ÷ 5   f 71 ÷ 10
g 49 ÷ 6   h 58 ÷ 8   i 48 ÷ 9

2 Farmer Jim has collected 58 eggs. He wants to put them into boxes of 6. How many boxes can he fill? How many eggs will be left over?

3 A group of 29 students want to get into two equal lines to play a game. Is this possible? Explain your answer.

4 Sindi has to fill a 60 litre tank with water. She has an 8 litre jug. How many jugfulls will she need to fill the tank?

you can use Workbook page 75

Rounding answers after division

There are 52 beads in the jar. I divide them among 5 people. How many beads does each person get?

52 ÷ 5 = 10 remainder 2

We cannot divide 2 beads among 10 people, so each person just gets 10 beads.

Divide. Round your answers.

1 47 among 3 groups

2 25 among 7 groups

3 89 among 9 groups

4 512 among 10 groups

5 88 among 5 groups

6 60 among 8 groups

Dividing by 10 and 100

Division is the inverse of multiplication. So, look what happens when you divide by 10 or 100.

30 ÷ 10 = 3

Thousands	Hundreds	Tens	Ones
		3	0
			3

1200 ÷ 100 = 12

Thousands	Hundreds	Tens	Ones
1	2	0	0
		1	2

- When you divide by ten the digits move one place to the right.
- When you divide by 100, the digits move two places to the right.

1 Try to do these divisions mentally.

a 40 ÷ 10   b 80 ÷ 10   c 110 ÷ 10
d 240 ÷ 10   e 350 ÷ 10   f 490 ÷ 10
g 520 ÷ 10   h 670 ÷ 10   i 990 ÷ 10

2 Try to do these divisions mentally.

a 400 ÷ 10   b 400 ÷ 100   c 4000 ÷ 100
d 800 ÷ 10   e 800 ÷ 100   f 8000 ÷ 100

3 Sal has 450 books. She puts an equal amount of books onto ten shelves. How many books on each shelf?

you can use Workbook page 76

Divide or multiply?

Remember division is the inverse of multiplication.
If you know that 5 × 9 = 45 then you can work out
that 45 ÷ 9 = 5 and 45 ÷ 5 = 9

1 For each multiplication write two related division facts.

a 14 × 9 = 126   b 15 × 6 = 90   c 14 × 8 = 112
d 19 × 3 = 57   e 18 × 7 = 126   f 9 × 13 = 117
g 8 × 11 = 88   h 16 × 5 = 80   i 4 × 17 = 68

2 Work out the missing number in each number sentence.

a 4 × ☐ = 48   b ☐ × 2 = 24   c ☐ × 3 = 36
d 3 × ☐ = 39   e 8 × ☐ = 48   f 5 × ☐ = 85

3 Joe buys big bags of sweets to sell on his stall.
He packs the sweets into smaller bags to sell them.

a For each bag below, work out how many bags he can fill.
Write how many sweets are left over.

| 89 | 93 | 180 | 95 | 75 |
| 6 in a bag | 8 in a bag | 10 in a bag | 7 in a bag | 4 in a bag |

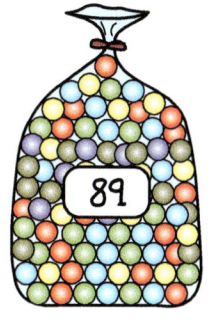

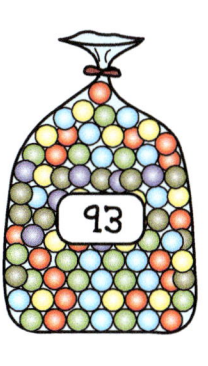

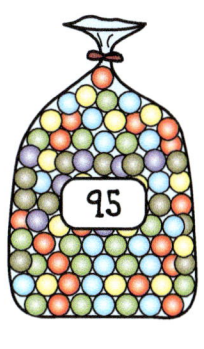

b How many sweets are left over altogether?

c Joe sells the leftover single sweets for 5c each. How much will he earn if he sells them all?

you can use Workbook page 77

Ratio and proportion

Look at this picture of a slice of bread.

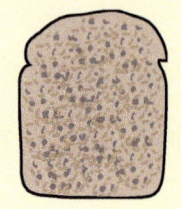

The picture is smaller than a real slice of bread. The real slice of bread is five times longer and five times wider than the bread in the picture.

You can work out the width and length of the real slice of bread using the measurements in the picture.

In the picture the bread is 20 mm wide and 25 mm high.

The real slice is 5 × 20 mm wide and 5 × 25 mm tall.

5 × 20 mm = 5 × 2 × 10 = 10 × 10 = 100 mm wide

5 × 25 mm = 5 × 20 + 5 × 5 = 100 + 25 = 125 mm tall

1 These animals are a fraction of the size of the real animals. Measure the length of each animal in the drawing and then work out the actual size of each animal.

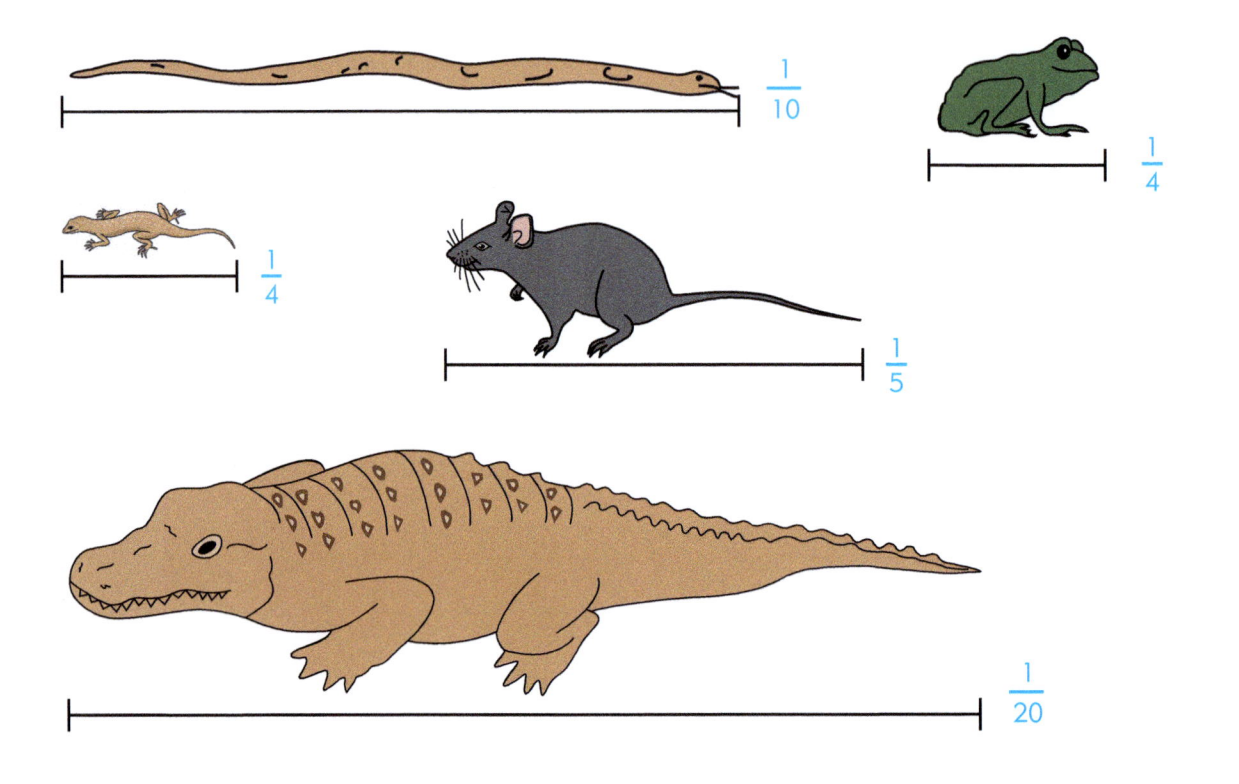

Ratio and proportion problems

1 Look at this picture of an ant.

  a The ant in the picture is 18 mm long. This is 4 times longer than the real ant. How long is the real ant?

  b In a group of ants there are 114 legs. How many ants in the group?

  c How many legs would there be if there were 100 ants?

2 Maria makes beaded friendship bangles. She needs 5 red beads, 10 blue beads and 8 yellow beads for each bangle.

  a She has 420 blue beads. How many bangles can she make?

  b She has 97 red beads. How many bangles can she make? Will she have any beads left over?

  c How many yellow beads will she need to make 32 bangles?

3 A bee has to visit about 4400 flowers to get enough pollen to make 10 grams of honey. How many flowers does the bee need to visit to get enough pollen for 1 gram of honey?

4 Aunty Zelma needs these ingredients to make 12 buns.

    1 cup flour   4 teaspoons baking powder
    $\frac{3}{4}$ cup milk $\frac{1}{2}$ cup sugar
    3 eggs   80 ml of melted butter

  Rewrite the ingredient list to show how much of each she would need if she wanted to make:

  a 24 buns   **b** 36 buns

5 This spider is three times the size of the real spider. How could you work out the length of the real spider?

Classroom proportions

Sanita drew this plan of her classroom. Each length on the plan is $\frac{1}{100}$th of the length of the real item.

This means that the real lengths are 100 times longer than those shown on the plan.

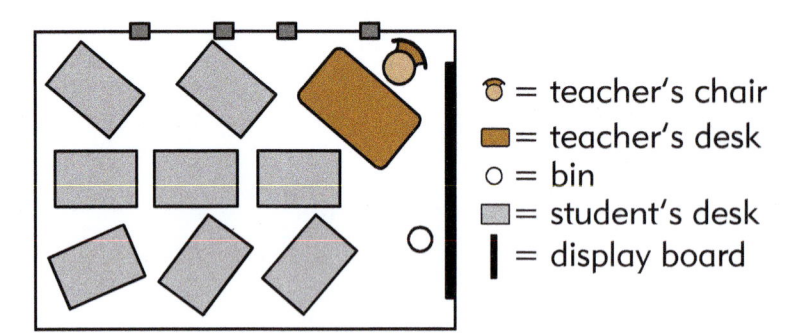

○ = teacher's chair
▬ = teacher's desk
○ = bin
▭ = student's desk
| = display board

1 Copy and complete this table to show the lengths of items on the plan and their size in the real classroom.

Item	Length on plan	Length in real classroom
Width of room		
Length of room		
Length of teacher's desk		
Length of display board		
Length of student's desk		
Width of student's desk		

2 Work with a partner. Discuss how you could draw a plan of your desk so that the length and the width on the plan are $\frac{1}{100}$th of the real length and width. Write down the steps you would follow.

you can use Workbook page 78